Andy
for your
library.

[illegible]

TISSUE CULTURE OF TREES

TISSUE CULTURE OF TREES

Edited by John H. Dodds

CROOM HELM
London & Canberra

AMERICAN EDITION
Published by
THE AVI PUBLISHING COMPANY, INC.
Westport, Connecticut 1983

Croom Helm Ltd, Provident House, Burrell Row,
Beckenham, Kent BR3 1AT

British Library Cataloguing in Publication Data

Tissue culture of trees.
1. Trees 2. Tissue culture
I. Dodds, John H.
634.9'56 QK477.2.T/

ISBN 0-7099-0830-X

Published in North, Central and South America 1983 by
AVI PUBLISHING COMPANY, 250 Post Road East,
Westport, Connecticut 06881

ISBN 0-87055-444-1

Printed and bound in Great Britain by
Biddles Ltd, Guildford and King's Lynn

CONTENTS

LIST OF CONTRIBUTORS

John H. Dodds (Editor)
Centro de Investigación y de Estudios Avanzados del IPN.
Unidad Irapuato,
Blvd. Gustavo Diaz Ordaz 237, Oficina 101.
Apdo. Postal 629,
Irapuato, Gto.
México 36660.

Teresa Bengochea
Centro de Investigación y de Estudios Avanzados del IPN.
Unidad Irapuato,
Blvd. Gustavo Diaz Ordaz 237, Oficina 101.
Apdo. Postal 629,
Irapuato, Gto.
México 36660.

Jennet Blake
Unit of Advanced Propagative Systems, Department of Horticulture,
Wye College,
University of London,
Nr. Ashford, Kent,
United Kingdom.

Allan John
Forestry Commission,
Northern Research Station,
Roslin,
Midlothian, Scotland,
United Kingdom.

Lorin W. Roberts
Dept. of Biological Sciences,
University of Idaho,
Moscow,
Idaho 83843,
USA.

Christopher P. Wilkins
Dept. of Plant Biology,
University of Birmingham,
PO Box 363,
Birmingham,
United Kingdom.

ACKNOWLEDGEMENTS

I would like to thank all the authors whose contributions have made this book possible. Their enthusiastic response to requests for chapters and their rigid adherence to the timetable have made my job as editor a very pleasant one. I must however acknowledge that I am responsible for any errors that may have crept in during the editing process.

I would like to thank Ms S. Newbury for assistance with line drawings and Miss J. Cox for photographic work. Much of the typing and modification of the manuscript was carried out by Margarita Huerta García to whom I am grateful. I would also like to thank the generosity of colleagues too many to name, for useful comments about the manuscript and for the provision of original plates and negatives.

Finally I would like to thank Mr T. Hardwick and the production team at Croom Helm Publishers for the helpful and efficient way in which they have handled production of this book.

LIST OF TERMS, ABBREVIATIONS AND SYNONYMS

ABA	Abscisic acid
Adenine	Aminopurine
Auxin	Plant growth regulator resembling IAA in physiological activity
Axenic	Aseptic
BAP	Benzylamino purine
Callus	Disorganised meristematic or tumour-like mass of plant cells
Cybrid	Cytoplasmic hybrid, heteroplast
Cytokinin	Plant growth regulator stimulating cell division and resembling kinetin in physiological activity. Mainly N_6 substituted aminopurine compounds
2,4-D	2,4-Dichlorophenoxyacetic acid
DNA	Deoxyribonucleic acid
Embryoid	Development of embryo-like structure under *in vitro* conditions; structure often has potential for further development into a plantlet
Explant	Excised fragment of plant tissue or organ used to initiate a tissue culture
GA_3	Gibberellic acid
Gibberellin	Plant growth regulator with gibbane ring exhibiting physiological activity similar to gibberellic acid
Glycine	Amino acetic acid
Haploid	Having a single set of chromosomes, monoploid
Heterokaryon	Fusion of unlike cells with dissimilar nuclei present
Heteroplast	Cell containing foreign organelles. Cytoplasmic hybrid, cybrid
Homokaryon	Fusion of similar cells
IAA	Indole-3 yl acetic acid
IPA	Isopentanyl adenine
Meristem culture	Apical meristem culture; explant consisting only of apical dome tissue distal to the youngest leaf primordium

Meristemoid	A cluster of meristematic cells within a callus capable of forming a primordium
MS	Murashige and Skoog (1962) medium
NAA	Naphthalene acetic acid
Nicotinic acid	Niacin
Organoid	Anomalous organ-like structure on culture
PEG	Polyethylene glycol
Shoot apex culture	Explant consisting of apical dome and a few subjacent leaf primordia
Synkaryote	Hybrid cell produced by fusion of nuclei in a heterokaryon

1 INTRODUCTION

John H. Dodds

The culture of fragments of plant tissue is not a particularly new science, in fact as long ago as 1893 Rechinger (1893) described the formation of callus on isolated fragments of stems and roots. The culture of plant tissues *in vitro* on a nutrient medium was performed by Haberlandt (1902), however, his attempts were unsuccessful because he chose too simple a medium that lacked critical growth factors.

Over the last fifty years there has been a surge of development in plant tissue culture techniques and a host of techniques are now available (Dodds and Roberts, 1982). The major areas are as follows.

Callus Culture

Callus is a rather ill-defined material, but is usually described as an unorganised proliferating mass of tissue. Although callus cultures have a great deal of potential in the biotechnological aspects of tissue culture, i.e. secondary product formation, they are not very suitable for plant propagation. The key reason for their unsuitability is that genetic aberrations occur during mitotic divisions in callus growth (D'amato, 1965). The aberrations can be of a major type, such as aneuploidy or endoreduplication. It follows therefore that the genetic status of the regenerated plants is different from that of the parent type. In general terms this genetic instability is undesirable, but there are occasions when a callus stage can be purposely included to diversify the genetic base of the crop.

A further problem with callus is that it loses its regenerative potential (Chandler, Dodds and Henshaw, 1982). By appropriate manipulation of the culture medium it is normally possible to induce the formation of roots or shoots (Skoog and Miller, 1957). If, however, the callus has already been through many subcultures, the ability to form these structures is lost. This loss of regenerative potential may be related to the accumulation of genetic aberrations.

Anther Culture

Haploid plants contain a single set of chromosomes which makes these plants of great use in breeding programmes. The phenotype is the expression of single copy genetic information, as there is no masking of a character because of gene dominance. Anther culture is used to produce haploid plants by inducing development of a single haploid pollen grain into a plantlet.

Tulecke (1953) observed that mature pollen grains of the gymnosperm *Ginkgo biloba* could be induced to form a haploid callus when transferred to a simple medium. In this example again we see the problem of a callus stage which allows genetic aberration to take place. Guha and Maheshwari (1966) cultured pollen grains of *Datura innoxia* and found that in the presence of either kinetin or coconut milk embryos were formed. Staining with acetocarmine stain showed that the plantlets produced were haploid (1967).

The use of anther and pollen culture therefore offers a useful tool to the conventional plant breeder as a way of obtaining haploid plants for selection of characters and inclusion in a breeding programme.

Meristem Culture

The term meristem culture has become a rather vague term and the use of words like 'meristemming' and 'mericloning' have been applied to sections of stem tips as long as 10 mm (Murashige, 1974). Cutter (1965) clearly distinguishes the apical meristem from the shoot apex. The apical meristem refers only to the region of the shoot apex lying distal to the youngest leaf primordium, whereas the shoot apex refers to the apical meristem plus a few leaf primordia. Explants with many leaf primordia are better referred to as shoot tips.

The uses of meristems and shoot tips in tissue culture are very varied and include (a) virus eradication, (b) micropropagation, and (c) storage of genetic resources, all of which are useful in the context of the subject of this book:

(a) *Virus eradication.* It is now over thirty years since Morel and Martin (1952) eradicated virus infections from *Dahlia* and *Solanum tuberosum* by the use of sterile meristem culture. The use of this technique is now standard practice with many important crop plants (Murashige, 1974b). Heat therapy (42°C) is often used in conjunction with meristem culture for virus eradication (Stace-Smith and Mellor,

1970). The size of the meristem explant is critical for virus eradication. Often so-called meristem tip cultures have failed to eliminate virus infection because the explant contained shoot apices with vascular tissue instead of true meristems.

(b) *Micropropagation.* The use of shoot apex cultures for the rapid clonal propagation of plant material was first carried out by Morel (1960). He cultured apices from orchids and found the production of many organised structures known as protocorms. If these protocorms were excised and transferred to fresh medium they would develop into whole plants. The application of this technique to the clonal propagation of trees has enormous potential. Techniques for the micropropagation of fruit trees have been developed in a number of laboratories (Jones and Hopgood, 1979; Wilkins and Dodds, 1982) and the rate of multiplication can be staggering. For example, the cherry rootstock 'Colt' can be micropropagated at such a rate that a single plantlet can be multiplied to give 4×10^{15} within one year.

(c) *Storage of genetic resources.* Many trees produce seed that is highly heterozygous in nature or that is recalcitrant. It follows therefore that seed is an unacceptable way of storing genetic resources from these tissues. Techniques are now being devised by which meristems of shoot tips from these plants can be stored *in vitro* and further details of this are given in Chapter 10.

Protoplast Technology

Isolated protoplasts are plant cells that have had the cell wall removed so that the plasmalemma is the only barrier between the living cytoplasm and the exterior environment. The use of protoplasts offers a wide range of new possibilities, for example the formation of new hybrids by fusion, or the insertion of new information by genetic engineering. Further details of these techniques are given in Chapter 9.

Despite the fact that tissue culture has been developing for about 50 years the range of plant material that has been studied is still very small. A great deal of work has concentrated on members of the *Solanaceae* family (i.e. Potato, Tomato, Tobacco, Henbane), as this family appears to be very 'plastic' in its developmental physiology. The Solanaceous plants are now used as model plants for the development of various culture techniques. In agricultural terms undoubtedly the most important crop plants are cereals and this is an area where development of tissue culture techniques has not yielded the success hoped for (Vasil, 1981).

The application of *in vitro* techniques to trees had a rather late beginning but is now becoming a very active area of research. However, it is again noticeable that only a relatively small number of species have been studied, and only a few techniques applied. Palms and some fruit trees have been studied in terms of micropropagation, but little work has been done in the area of protoplast technology or anther culture. One area that is particularly understudied is the application of micropropagation techniques to semi-tropical and tropical trees. This may well be for socioeconomic reasons rather than biological problems.

The application of these new techniques offers enormous economic potential in the long term and this pressure alone is likely to stimulate development of these areas.

It is impossible in a book of this type to cover all the trees that are being studied, many of which have great potential, but on which little work has been done (e.g. Jojoba). The book therefore contains articles which cover rather large areas (e.g. Gymnosperms), and also chapters on more specific areas, such as application of certain technologies to trees in general. It is hoped that the description of the potential technologies together with the specific examples cited in certain chapters will stimulate further research into *in vitro* culture of trees, an area of research that is still in its infancy.

References

Chandler, S.F., Dodds, J.H. and Henshaw, G.G. (1982) 'Factors affecting adventitious shoot formation in *Solanum laciniatium* callus cultures'. *Proceedings 5th International Plant Tissue Culture Congress*. Japan 1982 (in press)

Cutter, E.G. (1965) 'Recent experimental studies of the shoot apex and shoot morphogenesis'. *Bot. Rev. 31*, 7-113

D'amato, F. (1965) 'Endoploidy as a factor in plant tissue development' in P.R. White (ed.), *Proceedings International Conference Plant Tissue Culture*, pp. 449-62

Dodds, J.H. and Roberts, L.W. (1982) *Experiments in Plant Tissue Culture*, University Press, Cambridge, p. 200

Guha, S. and Maheshwari, S.C. (1966) 'Cell division and differentiation of embryos in the pollen grains of *Datura in vitro*'. *Nature 212*, 97-8

Guha, S. and Maheshwari, S.C. (1967) 'Development of embryoids from pollen grains of *Datura in vitro*'. *Phytomorphology 17*, 454-61

Haberlandt, G. (1902) 'Kulturversuche mit isolierten Pflanzenzellen'. *Sber. Akad. Wiss. Wein 111*, 69-92

Jones, O.P. and Hopgood, M.E. (1979) 'The successful propagation *in vitro* of two rootstocks of *Prunus*: the plum rootstock Pixy (*P. insititia*) and the cherry rootstock F1311 (*P. avium*)'. *J. Hort. Sci. 54*, 63-6

Morel, G. (1960) 'Producing virus-free Cymbidiums'. *Am. Orchid Soc. Bull. 29*, 495-7

Morel, G. and Martin, G. (1952) 'Guérison de dahlias alteistes d'une maladie à virus'. *Compt. Rend. Acad. Sci. 235*, 1324-5
Murashige, T. (1974a) 'Plant propagation through tissue culture'. *Ann. Rev. Plant Physiol. 25*, 135-66
Murashige, T. (1974b) 'Plant cell and organ culture methods in the establishment of pathogen free stock'. *A.W. Dimock Lecture*, Cornell, Ithaca, N.Y. p. 26
Rechinger, C. (1893) 'Untersuchungen über die Grenzen der teilbarkeit im Pflanzenreich'. *Abh. zool. bot. Ges. 43*, 310-34
Skoog, F. and Miller, C.O. (1957) 'Chemical regulation of growth and organ formation in plant tissues cultured *in vitro*'. *Symp. Soc. Exp. Biol. 11*, 118-30
Stace-Smith, R. and Melor, F.C. (1970) 'Eradication of potato spindle tuber virus by thermotherapy and axillary bud culture'. *Phytopathology 60*, 1857-8
Tulecke, W. (1953) 'A tissue derived from pollen of *Ginkgo biloba*'. *Science 117*, 599-600
Vasil, I.K. (1981) 'Cereal crop improvement'. *IAPTC newsletter*
Wilkins, C.P. and Dodds, J.H. (1982) 'The application of tissue culture techniques to problems of plant genetic conservation'. *Science Progress* (in press)
Zenk, M.H. (1978) 'The impact of plant cell culture on industry' in T.A. Thorpe (ed.), *Frontiers of plant tissue culture 1978*, IAPTC Calgary, Canada

2 TISSUE CULTURE OF CONIFEROUS TREES

Allan John

Conifers or softwoods are general terms that are applied to members of the Gymnospermae, a group that is subdivided into Ginkogaceae, Taxaceae and Pinaceae (Dallimore and Jackson, 1954). Most economically important species are grouped in Pinaceae, which includes the Cupressineae (Cypress tribe), the Taxodineae (*Cryptomeria* and *Sequoia*) and the Abietineae (Fir tribe) which contains the important genera *Pinus*, *Picea*, *Abies*, *Pseudotsuga*, *Cedrus* and *Larix*.

The group has a distribution that stretches from the sub-arctic to the tropical regions of the world. Some of the species have a very large latitudinal spread, *Picea sitchensis* being found from Alaska to Northern California, whereas others have a narrow geographical distribution, *Pinus radiata* being naturally restricted to the Monterey Peninsula of California.

Many countries, such as the USA and USSR, have used their indigenous softwood forests for the production of timber, paper pulp and other products, but these natural resources have either long been exhausted in many other countries or did not exist there. Conifers have been introduced into some of these countries as exotic species and many trials and selections have occurred to find the particular species most suitable to the requirements of each country. Hence *Picea sitchensis* and *Pinus contorta* are grown in the UK, *Pseudotsuga menziesii* is grown in continental Europe and *Pinus radiata* is grown in New Zealand, these all being species originating from North America. Indeed, within a species, individuals from certain areas, or origins within its geographical range, have been found to be more suitable; *Picea sitchensis*, from the Queen Charlotte Islands in Canada, being grown in the United Kingdom is a good example.

Forest planting stock production in the past has relied on, and to a large extent still relies on, the use of seed collected from natural stands. However outstanding individuals have been selected in mature plantations and have been used as the basis for breeding programmes, with progeny trials to test the best genetic combinations and seed orchards established to realise the potential gains. However, seeds from seed orchards are at best open pollinated with only one parent known.

Techniques exist to make controlled genetic crosses, but they are expensive. Vegetative propagation can be seen as a method of multiplying selected genotypes on a large scale and extensive studies have been made to investigate the potential use of stem cuttings in this context (Kleinschmidt, 1974). *In vitro* (sterile axenic culture) methods of vegetative propagation are also being studied by many workers with a view to their use in future commercial forestry.

Softwood species were among the first to be used for *in vitro* studies. Schmidt (1924) cultured and observed the growth and development of embryos of various *Pinus* species and Gautheret (1934) described the development and cytology of callus formed from cambial explants of *Pinus pinaster*. Most of the work reported subsequently was on embryo (La Rue, 1935; Loo and Wang, 1943; Sterling, 1949) or root culture (Slankis, 1947). The present interest in the *in vitro* culture of conifers was encouraged by Ball (1950) who reported that callus of *Sequoia sempervirens* could be subcultured and that morphogenesis could be induced on its surface. Many researchers realised that this developed the *in vitro* culture of conifers from being an interesting laboratory phenomenon to a technique that had applications in many areas of forestry.

The *in vitro* culture of conifers has been reviewed extensively in recent years (Winton and Huhtinen, 1976; Bonga, 1977; Mott, 1978; David and Thomas, 1979; Mott, 1981; David, 1982). An attempt will be made here to review some of the relevant work to date and to put it in the context of commercial forestry and its specialised requirements.

Explants for *in vitro* culture have been derived from almost every tissue present during the life history of conifers: male and female gametophytic tissue; immature and mature embryos; hypocotyl, cotyledons and shoot apex of newly germinated seedlings; apical meristems; whole shoots; juvenile and mature needles; brachyblasts (needle fascicles) and various cambia in more mature shoots. These tissues have been subjected to treatments that induce callus formation, induce dividing cells to form suspension cultures, induce adventitious bud formation or promote axillary bud development.

Callus Culture

Callus cultures have generally been derived from juvenile tissues and there does not seem to be a general methodology for callus formation and subsequent morphogenesis. Cheng (1975) initiated callus from *Pseudotsuga menziesii* cotyledon tissue with a medium containing absisic acid (ABA), indolebutyric acid (IBA), benzylamino purine (BAP) and 2-isopentyl purine (2iP) and found that the initiation of shoot meristems on the callus was promoted by adding a solution containing a relatively high concentration of BAP (0.5–1.0 mmol) to the medium. However Reilly and Brown (1976) obtained callus from immature embryos of *Pinus radiata* on a medium containing auxin and cytokinin and found that morphogenesis occurred when the calluses were transferred to a hormone-free medium. John (unpublished data) observed a similar phenomenon in friable callus obtained from mature embryos of *Picea sitchensis* and *Pinus contorta*.

The methods of initiation of callus and subsequent induction of morphogenesis seem to be specific to the species being investigated. The approach to callus initiation has necessarily been empirical and most workers have tried numerous combinations of growth substances and physical environments to achieve the desired results. This has not always been successful. Minocha (1980) exposed callus derived from embryos of *Pinus strobus* to over 100 combinations of auxin and cytokinin without observing morphogenesis and shoot formation. Even when a standard technique has been evolved, morphogenesis in established calluses is not always reliable. John (unpublished data) observed that transfer of *Picea sitchensis* callus from hormone-containing to hormone-free media did not always result in morphogenesis. This result was observed even in callus from full sib embryos (genotypes with the same parents) where the narrow genetic base would be expected to remove the variation.

David (1982) has argued that, unlike some herbaceous angiosperm calluses, the full morphogenetic potential of calluses derived from gymnnosperms has not yet been fully realised and that callus culture should not be dismissed as a method of producing large numbers of propagules. He further states that the morphogenetic potential of calluses already achieved should be investigated to determine whether it can be maintained indefinitely or whether it diminishes with each subculture.

It must also be shown whether callus cultures of gymnosperms are genetically stable and if the propagules produced from them are true to type. Patel and Berlyn (1981) reported that cells from callus and regenerated buds of *Pinus coulteri* have an increased DNA level and increased chromosomal aberrations.

Suspension Culture

Suspensions of cells derived from calluses offer by far the greatest potential for multiplication of selected genotypes, presuming that cells can be kept in an actively dividing state in culture and that embryogenesis and development of the induced embryoids to whole plants can be achieved at will. Total success has not been achieved to date but limited successes in some steps of the process have been reported. Reinert (1956) showed that if vitamin B12 and folic acid were withheld from calluses of *Picea glauca*, the callus changed from a slow-growing firm structure to a rapidly growing gelatinous callus that dissociated into tissue fragments and single cells in liquid culture. Durzan and Stewart (1968) obtained single cell suspension cultures of *Picea glauca* and *Pinus banksiana* that underwent cell division and growth. Chalupa and Durzan (1973) obtained suspension cultures of *Picea abies* and found that after repeated transfers to a fresh medium, some small embryoids of 5-10 cells were formed. Winton and Verhagen (1977) initiated embryoids of 3-5 cells in suspension cultures derived from callus of *Pseudotsuga menziesii* and *Pinus taeda*. However the authors did not prove to their own satisfaction that true embryoids were formed, as no subsequent development into plantlets occurred. More recently, Durzan (1979) has reported the establishment of suspension cultures from shoot tips of 3-4-year-old *Pseudotsuga menziesii*. Globular embryos that became polarised along a root shoot axis with cotyledon-like structures developed in the suspensions.

The results reported above and elsewhere are encouraging and it seems only a matter of time before the major problem of promoting plantlet formation from the initiated embryoids is solved. It would then be possible to produce almost infinite clonal replicates of any chosen genotype of any desired species. However, as in callus cultures, it would have to be carefully established that the genotype remains stable through the various culture processes and that the plantlets produced are genetically identical to the plant from which the original explant was taken.

Bud Induction

The problems of uncertain genetic stability in callus and suspension cultures have prompted many workers to look for regions of meristematic activity in intact plants that could be used as sources of material. David (1982) has stated that these regions are resting primary meristematic cells located in the axils of cotyledons and juvenile leaves, or at the apex of brachyblasts, and structures developed from these regions should be termed axillary buds or shoots. It is also possible to induce the formation of buds and shoots directly on existing plant structures from tissues that were never destined for that fate and the morphogenesis that occurs in this situation should be termed adventitious.

Adventitious Bud Initiation

Adventitious buds have been induced to form on intact embryos, excised cotyledons, hypocotyl segments, needles and shoot apices. The explant tissue has been juvenile in nearly all cases.

Embryos and Cotyledons. The mixture and concentration of growth substances used to induce adventitious bud formation has been shown to be different for the various species investigated. BAP has been found to be the best cytokinin for response in a number of instances, either alone (Sommer, 1975; Winton and Verhargen, 1977; Reilly and Washer, 1977) or combined with one or more auxins (Sommer and Brown, 1974; Chalupa, 1977; Cheng, 1976). Other cytokinins, such as 2iP or kinetin, have been shown to be less effective (Cheng, 1977; Webb and Street, 1977).

The buds appeared to arise from the epidermal or subepidermal layers of the tissue used (Mott, Smeltzer and Mehra-Palta, 1977; Cheah and Cheng, 1978). Subepidermal cells of *Pseudotsuga menziesii* cotyledons began division after 4 days in culture and bud primordia were visible after 20 days (Cheah and Cheng, 1978). Mott (1981) reported that buds formed on the surface of excised *Pinus taeda* cotyledons after 6-10 weeks in culture and extended to 5-10 mm whilst still connected to the explant.

The presence of growth substances in the medium initiated the formation of adventitious structures but usually prevented their development into shoots. Removal of the hormonal influence by transfer of the explants plus adventitious buds to a hormone-free medium was necessary before elongation to form shoots occurred (Campbell and Durzan, 1975; Cheng and Voqui, 1977; Winton and Verhagen, 1977). In some cases, excision of the induced adventitious buds from the explant and transfer to a hormone-free medium was necessary before shoot formation occurred (Coleman and Thorpe, 1977; Reilly and Washer, 1977; Mehra-Palta, Smeltzer and Mott, 1978; Von Arnold and Eriksson, 1981). Mott (1981) has stated that the difficulties in achieving shoot elongation in the induced adventitious buds have resulted in a lack of rooted clones for field testing and has postulated that the conditions of culture could have initiated some form of dormancy.

Hypocotyl Segments. The hypocotyl is a specialised structure that, in conifers, extends to force the cotyledons above the soil level as the seed germinates. The hypocotyl is the first stem unit of the conifer to form (Esau, 1965).

Adventitious buds have been initiated on excised hypocotyl segments of a number of species. BAP has been found to be effective in *Picea glauca* (Campbell and Durzan, 1975, 1976), *Picea abies* (Chalupa, 1975) and *Biota orientalis* (Thomas, Duhoux and Vazart, 1977). Adventitious bud initiation has been shown in response to a medium low in organic nutrients, and the rate of initiation was increased by the addition of growth substances such as ABA or BAP plus NAA (naphthalene acetic acid) or BAP alone (Isikawa, 1975). Isikawa (1978) demonstrated that the morphactin chlorflurenol promoted the formation of abnormal adventitious buds on hypocotyls of *Cryptomeria japonica*.

Campbell and Durzan (1975) suggested that the adventitious buds arose from the superficial tissues of the hypocotyl explants, whereas Vazart, Conciecao and Thomas (1979) reported that morphogenesis was initiated in the outer cortical tissues from cells that had remained mitotic.

The presence of growth substances in the bud induction medium inhibited the development of the induced adventitious buds. Shoots could be obtained if the buds, with some of the hypocotyl explant still attached, were transferred to a hormone-free medium (Campbell and Durzan, 1975; Chalupa, 1977).

Needles. The induction of adventitious buds directly on needles would be a preferred technique, since, in terms of clonal propagation, many explants could be achieved per clone compared to the one achieved by using an intact embryo as the explant source.

Jansson and Bornman (1980) obtained needle explants from the flushing zone of 20-22-week-old *Picea abies* seedlings that had been physiologically aged to one year by a frost hardening treatment. Two to three adventitious buds were formed on 22 per cent of the needles with 5 μmol BAP and 50 μmol NAA and outgrowth of the buds to form shoots was achieved by transfer to a hormone-free medium containing one-third strength mineral nutrient. Jansson and Bornman (1981) found bud initiation in needle explants was greatly affected by age, as defined by needle length. Five millimetre needles produced buds all over their surface, 10 mm needles produced buds proximal and immediately distal to the abscission zone and 15 mm needles produced buds proximal to the abscission zone only. Buds proximal to the abscission zone were initiated from epidermal cells and from subsidiary cells of the stomata, whereas those distal to the abscission zone were initiated from epidermal cells only. Von Arnold and Eriksson (1979) obtained similar results with 5-year-old *Picea abies* needle explants collected just after bud flush. They also demonstrated that adventitious bud yield could be improved by incubation of the explants *in vitro* at 10-16°C before transfer to 20°C. The capacity of the needles to form adventitious buds varied from about 10 per cent to 60 per cent, depending on the clone. Shoot extension was achieved by transfer to a hormone-free medium. Von Arnold and Eriksson (1979) also report preliminary results suggesting that the technique might be applied to more mature trees.

The technique described by Reilly and Brown (1976) did not utilise adventitious bud formation directly on the needle explant but included a callus stage. Primary needle segments from 10-12-week-old *Pinus radiata*, germinated *in vitro*, were treated with BAP which resulted in renewed cell division and the production of meristematic areas near the cell surface. A smooth compact callus formed along portions of the outer surface of the needle. Transfer of the needle segments to a medium containing half-strength nutrients and no growth substances resulted in the formation of small apical meristems that could be induced to form shoots by excision and transfer to new hormone-free medium.

Shoot Apices. Chalupa (1977b) cultured excised dormant vegetative buds of *Picea abies* and *Pseudotsuga menziesii in vitro* and found that a low concentration of auxin (0.01-0.3 mg/l) in the medium stimulated elongation. Webb and Street (1977) initiated callus from vegetative buds of *Pinus contorta* and *Picea sitchensis* and induced morphogenesis on the callus. Tranvan (1979) attempted to induce adventitious bud formation at the apex of very young *Pinus sylvestris* seedlings, but, although multi-budded apices were formed, it was not clear whether bud initiation was occurring directly at the seedling apex, on the cotyledons or on internodal tissue.

Von Arnold and Eriksson (1979b) have achieved adventitious bud formation on vegetative buds of *Picea abies* collected just after flushing from trees that varied in age from 6 to 50 years. The needle primordia of the excised buds became irregularly shaped in response to BAP in the medium. The bud-like structures that formed after 8 weeks in culture were induced to form shoots by transfer to a hormone-free medium. There was no apparent correlation between the age of the donor tree and adventitious bud formation. Anatomical investigations showed that adventitious bud initiation was *de novo* from needle primordia and not from axillary bud primordia present in the vegetative buds.

Axillary Bud Initiation

It has been suggested that axillary bud initiation can be promoted by growth substances added to the medium and generally BAP at between 10^{-5} and 5×10^{-5} M with an auxin (NAA or IBA) at 10^{-8} M, applied over a period, have been suggested as the most successful concentrations (David, 1982). David, Isemukali and David (1978) induced axillary buds on explants of *Pinus pinaster* hypocotyl sections with BAP and NAA.

Mapes, Young and Zaerr (1981) have described a method for inducing axillary buds in the axils of cotyledons of *Pseudotsuga menziesii* and they found that cytokinin and auxin were a prerequisite for development. However, extension of the axillary buds was only achieved when the bud clusters were transferred to a hormone-free medium.

The presence of growth substances, particularly auxin, in the medium can give rise to callus development on the explant and it can be difficult to distinguish between axillary and adventitious bud initiation when it occurs in the region of callus development (John, unpublished observations).

The presence of cytokinin in the medium seems to be necessary for axillary bud development in many herbaceous species (Hussey, 1980). This is not necessarily the case for some conifer species. Boulay (1979) developed a system for the *in vitro* multiplication of *Pseudotsuga menziesii* which was not dependent upon growth substances in the medium for axillary bud development. *Pseudotsuga menziesii* seedlings germinated *in vitro* were grown on a culture-establishment medium. The shoots were subdivided and transferred to a medium with a reduced level of ammonium nitrate, which resulted in axillary bud outgrowth. The axillary buds were excised and transferred to a medium containing activated charcoal that promoted extension. At the end of the extension phase the shoots were subdivided and returned to the multiplication medium.

Some of the systems described for the initiation and outgrowth of axillary buds are very complicated and require a number of medium changes. However, John and Murray (in press) have described a system for the proliferation of *Picea sitchensis* on a single medium without growth substances. Seedlings at the cotyledon stage were surface sterilised, put into culture, and the axillary buds and shoots that developed were subcultured to fresh medium. The subcultures produced axillary buds and shoots during their extension. The authors noted however that there was a difference in the rate of bud production in the different types of subculture, i.e. although the subcultures extended at the same rate, those derived from apical buds produced more axillary buds and shoots per unit length of stem than those derived from axillary buds. However when the apex of an axillary bud subculture was transferred through a number of subcultures, its rate of axillary bud and shoot formation increased to levels comparable with those of the original apices.

Rooting

In vitro techniques produce vegetative shoots that are generally devoid of a root system. There are two possible ways of establishing roots: *in vitro* or by the techniques that have been developed for rooting conifer cuttings under intermittent mist.

Campbell and Durzan (1976) transferred excised *Picea glauca* adventitious shoots to a hormone-free medium for rooting and the rooted shoots grew vigorously in soil. Brown and Sommer (1977) found that rooting *in vitro* was sporadic and Von Arnold and Eriksson (1981) showed that spontaneous rooting of adventitious shoots of *Pinus contorta* was enhanced by auxin. Slow growing shoots of *Pinus radiata* did not root *in vitro* whereas rapidly extending shoots did, without auxin treatment (Reilly and Washer, 1977).

The use of plant growth substances is an established technique for promoting the rooting of cuttings of a vast number of species in both horticulture and forestry. Mott *et al.* (1977) rooted adventitious shoots of *Pinus taeda* on a medium containing 0.1 mg/l IBA and NAA. Other workers have obtained 50 per cent rooting of *Araucaria cunninghamii* axillary shoots on a medium containing 6 μmol IBA, whereas Winton and Verhagen (1977) achieved less than 5 per cent rooting on adventitious shoots from *Pseudotsuga menziesii* in response to 10 μg/l IBA.

The rooting process can be separated into two stages: root initiation and root outgrowth. They have been found to occur in response to different media. Sommer, Brown and Kormanik (1975) showed that adventitious shoots of *Pinus palastris* rooted spontaneously on a hormone-free medium, but four weeks pretreatment with auxin increased rooting. Mehra-Palta *et al.* (1977) found that for optimum rooting, adventitious shoots from *Pinus taeda* must be extending and longer than 5 mm with a recognisable stem portion. Roots were initiated on the shoots after five weeks in a medium containing 0.05 mg/l BAP and 0.1 mg/l NAA and root outgrowth was then achieved by transfer to a hormone-free medium, with a rooting level of 50 per cent. They found that a similar level of rooting was possible after 12 weeks under mist and that the roots formed under non-sterile conditions were more vigorous. Cheng and Voqui (1977) have reported a similar rooting system for adventitious shoots of *Pseudotsuga menziesii.* High levels of rooting (about 80 per cent) were achieved by transfer to a hormone-free medium after four weeks initiation on a medium low in sucrose (0.5 per cent) with 0.25 μmol NAA. Survival was high, about 90 per cent, when the rooted shoots were transferred to soil from *in vitro* culture. Mott and Amerson (1981) have suggested that the survival of rooted *Pinus taeda* shoots in soil depended on the roots being at least 3 mm long and a high relative humidity being maintained.

Rooting levels *in vitro* are often low and frequent media changes are difficult and expensive. This has prompted some workers to attempt rooting directly under mist, i.e. under non-sterile conditions. Webb and Street (1977) achieved 20 per cent rooting of *Picea sitchensis* and *Pinus contorta* shoots under mist and found that adventitious shoots from callus initiated with 2ip or kinetin rooted better than those initiated with other cytokinins. David *et al.* (1978) basally soaked *Pinus pinaster* adventitious shoots in a commercial rooting solution for 24 hours and 90 per cent of the shoots rooted after one month in a peat-perlite mixture. A 24-hour basal soak of *Sequoia sempervirens* shoots in IBA resulted in 80 per cent rooting in a perlite-vermiculite mixture after 30 days, and the more vigorous shoots rooted best (Boulay, 1979).

Horgan and Aitkin (1981) have described a technique that used both *in vitro* and non-sterile methods. Adventitious shoots of *Pinus radiata* were pretreated for 5 days on a medium containing 2 mg/l IBA and 0.5 mg/l NAA. The pretreated shoots were transferred to a peat-pumice medium and rooting occurred in a humid environment with occasional misting. The rooting level was 86 per cent after 42 days compared to 36 per cent with no hormone pretreatment.

Application of in vitro Techniques to Conifer Plant Production

Conifers are generally produced from seed that is germinated in prepared seedbeds and the seedlings grown on to a size suitable for forest planting. Silvicultural and nursery techniques have been refined over the years to such a degree that plantable trees are very cheap to produce, for example *Picea sitchensis* seedlings produced by commercial nurseries in the UK cost approximately £0.05 sterling each in 1981. Methods are being evolved for the multiplication of selected genotypes by stem cuttings. After rooting and growth, further multiplication can be achieved by taking more stem cuttings from the already rooted cuttings. It is estimated that, using this and similar techniques, clonal multiplication rates of between 250 and 1000 might be achieved (P. Biggin, unpublished data). The genotype selected as the explant source for *in vitro* propagation must be superior, or at least improved, since the increased growth of the genotype and its subsequent increased yield of products will have to pay for the increased cost of the method of production of the planting stock. It has been estimated that rooted plantlets produced by *in vitro* methods might cost 3-30 times as much as seedlings produced using normal nursery techniques (Brown and Sommer, unpublished data, from Sommer and Caldas, 1981). Rediske (1978) is much less pessimistic and estimates that the cost of *in vitro* plants should be significantly lower than that of rooted stem cuttings which are estimated to cost at least 30 per cent more than seedlings (Kleinschmidt, 1974).

The potential genetic gain from tree breeding programmes is not realised for some time, since seed orchards must be established and seed production does not reach a maximum for a number of years. McKeand (1981) has suggested that the commercial production of *in vitro* plantlets could be possible one or two years after selection, compared to 10-15 years for the production of commercial quantities of improved seed from seed orchards of *Pinus taeda*.

There are many ways in which the potential genetic gain might be realised with *in vitro* techniques. Tissue culture explants could be taken from seeds or young seedlings. The plant material would be untested and the estimated genetic gains would be statistically based on previous progeny trials, since it is almost impossible to correlate the growth and development of juvenile material with its subsequent performance as a mature tree (McKeand, 1981). Explants from mature, tested genotypes could be used and Von Arnold and Eriksson (1978b) have successfully induced buds on vegetative buds of mature *Picea abies* but no information is available yet on the performance of the plants in the field. Explants could be obtained from tested mature material that had been rejuvenated and the rooted plantlets formed would show all the characteristics of juvenility that are necessary for the normal growth and development of the tree. Explants could also be obtained from material that had been cryopreserved in a juvenile state whilst the genotype was being tested under field conditions. Cold storage of plant material for up to 15 years could be necessary if this technique was applied.

Phase change, within the tree, from juvenile to mature characteristics, will always present problems to the vegetative propagator. Plant material which can be propagated successfully is juvenile and untested, while mature material that is tested cannot be successfully propagated. Bonga (1982) has argued that the rejuvenation of cells and tissues is a very important feature in achieving the effective cloning of forest trees. Techniques of sequential grafting are being evolved to rejuvenate mature conifers. Franclet (1980) grafted scions from a 75-year-old Douglas fir tree onto juvenile rootstocks and then regrafted the scions, after flushing, onto further juvenile rootstocks. By the fifth graft, apical vigour had increased and was similar to that of autografts of 8-month-old seedlings. John and Fletcher (unpublished data) found an increase in the rooting ability of mature *Picea sitchensis* cuttings after one graft cycle. 'Rejuvenated' shoots produced by the sequential grafting technique show juvenile characteristics *in vitro* (Franclet, 1979).

The production of propagules by *in vitro* techniques must be economic. Some of the propagation methods that have been described are very complex and some relatively simple. A very complex example is:

(1) Induction of adventitious buds on an excised, intact embryo on a medium containing growth substances.
(2) Outgrowth of the induced adventitious buds into shoots on a growth-substance-free medium.
(3) Excision of shoots and root initiation on a medium containing growth substances.
(4) Outgrowth of adventitious roots on a medium without growth substances.
(5) Plant establishment in soil.
(6) Forest planting.

A more simple example is:

(1) Extension of a surface sterilised shoot and the proliferation of axillary buds on a medium without growth substances.
(2) Excision of the axillary shoots and further extension and proliferation of axillary buds on a medium without growth substances.
(3) Adventitious rooting under intermittent mist.
(4) Plant establishment in soil.
(5) Forest planting.

Each additional handling stage introduces further expense into plant production. It is possible that some species might never be propagated economically *in vitro* if the cost of the technique could not be reduced to realistic levels. Mechanisation would reduce costs, but the techniques are not likely to be available in the foreseeable future.

Techniques that induce adventitious buds on embryo, hypocotyl and cotyledon explants are limited since the explant is destroyed in the propagation process. The method is extremely wasteful of plant material that might have been expensively produced by controlled pollinations, and the rates of propagation are fairly low. Aitken-Christie (personal communication) has estimated a multiplication rate of 124 per clone in 17 months by adventitious bud initiation on excised cotyledons of *Pinus radiata*. Clones could be preserved in culture if adventitious buds were induced to form on the needles of the induced adventitious buds using the techniques of Janssen and Bornman (1980), but there are no reports of this having been tried. Suspension and subculturable callus cultures maintain the clone *in vitro*. Propagation rates would be extremely high and embryoids or adventitious shoots could be harvested continually from them. However the problems of uncertain gene stability must cast doubts on these methods. Needle and vegetative bud explants can be taken repeatedly since the donor clone can be maintained under non-sterile conditions. Adventitious shoot initiation is seasonal to some degree in that the technique requires the explants to be in specific physiological and morphological conditions. Year-round shoot production might be possible only by expensive physiological manipulations of the stock trees in controlled environments. *In vitro* propagation by axillary bud initiation offers many advantages although it is limited to species that freely form axillary buds in culture. The shoots can be maintained through numerous subcultures, for up to 2 years, without apparent ageing (John, unpublished data), there is no callus stage to introduce genetic aberrations, the clone is maintained in culture and the number of shoots increases geometrically with each subculture.

New technologies are being developed which will have direct application to forestry. Genetic manipulation will create new genotypes, perhaps even nitrogen-fixing trees. Protoplast culture will introduce new variation into the tree populations by mutation and new species and hybrids will be created by somatic hybridisation. Dihaploids will be formed, by chromosome doubling in cells derived from haploid tissue, that will give the opportunity to conifer tree breeders to produce homozygous inbred lines. These are techniques for the future. The *in vitro* culture of conifers is mainly involved with propagation at present and when the various techniques have been refined and are in use, the new technologies will be introduced.

References

Ball, E. (1950) 'Differentiation in a Callus of *Sequoia sempervirons*', *Growth, 14*, 295-325

Bonga, J.M. (1977) 'Applications of Tissue Culture in Forestry' in J. Reinert and Y.P.S. Bajaj (eds.), *Plant Cell and Organ Culture*, Springer-Verlag, Berlin, pp. 93-108

Bonga, J.M. (1982) 'Vegetative Propagation in Relation to Juvenility, Maturity and Rejuvenation' in J.M. Bonga and D.J. Durzan (eds.), *Tissue Culture in Forestry*, Nijhoff and Junk, The Hague, pp. 387-412

Brown, C.L. and Sommer, H.E. (1977) 'Bud and Root Differentiation in Conifer Cultures', *Tappi, 60*, 72-3

Boulay, M. (1979) 'Propagation *In Vitro* du Douglas par Micropropagation de Germination Aseptique et Culture de Bourgeons Dormant', *Etudes Rech. AFOCEL, 12*, 67-75

Campbell, R.A. and Durzan, D.J. (1975) 'Induction of Multiple Buds and Needles in Tissue Cultures of *Picea glauca*', *Can. J. Bot., 53*, 1652-7

Campbell, R.A. and Durzan, P.J. (1976) 'Vegetative Propagation of *Picea glauca* by Tissue Culture', *Can. J. Forest Res., 6*, 240-3

Chalupa, V. (1975) 'Induction of Organogenesis in Forest Tree Tissue Cultures', *Communicationes Instituti Forestalis Cechosloveniae, 9*, 39-50

Chalupa, V. (1977a) 'Organogenesis in Norway spruce and Douglas Fir Tissue Cultures', *Communicationes Instituti Forestalis Cechosloveniae, 10*, 79-87

Chalupa, V. (1977b) 'Development of Isolated Norway Spruce and Douglas Fir Buds *In Vitro*', *Communicationes Instituti Forestalis Cechosloveniae, 10*, 71-8

Chalupa, V. and Durzan, D.J. (1973) 'Growth of Norway Spruce (*Picea abies*) Tissue and Cell Cultures', *Communicationes Instituti Forestalis Cechosloveniae, 8*, 111-25

Cheah, K.T. and Cheng, T.Y. (1978) 'Histological Analysis of Adventitious Bud Formation in Cultured Douglas Fir Cotyledons', *Am. J. Bot., 65*, 845-9

Cheng, T.Y. (1975) 'Adventitious Bud Formation in Culture of Douglas Fir (*Pseudotsuga menziesii* Mirb. Franco)', *Plant Sci. Lett., 5*, 97-103

Cheng, T.Y. (1976) 'Vegetative Propagation of Western Hemlock (*Tsuga heterophylla*) through Tissue Culture', *Plant Cell Physiol., 17*, 1347-50

Cheng, T.Y. (1977) 'Factors Affecting Adventitious Bud Formation in Cotyledon Culture of Douglas Fir', *Plant Sci. Lett., 9*, 179-87

Cheng, T.Y. and Voqui, T.H. (1977) 'Regeneration of Douglas Fir Plantlets through Tissue Culture', *Science, 198*, 306-7

Coleman, W.K. and Thorpe, T.A. (1977) '*In vitro* Culture of Western Red Cedar (*Thuja plicata* Donn.) 1. Plantlet Formation', *Bot. Gaz., 138*, 298-304

Dallimore, W. and Jackson, A.B. (1954) *A Handbook of Coniferae*, Edward Arnold, London

David, A. (1982) 'In Vitro Propagation of Gymnosperms' in J.M. Bonga and D.J. Durzan (eds.), *Tissue Culture in Forestry*, Nijhoff and Junk, The Hague, pp. 72-108

David, A. and Thomas, M.J. (1979) 'Organogenèse et Multiplication Végétative *In Vitro* Chez les Gymnospermes', *Année Biologique 18*, 381-416

David, H., Isemukali, K. and David, A. (1978) 'Obtention de Plants de Pin Maritime (*Pinus pinaster* sol.) à Partir de Brachyblasts au d'Apex Caulinaires de Très Jeunes Sujets Cultivés *in vitro*', *Compte. Rendu. Acad. Sci., 287*, 245-8

Durzan, D.J. (1979) 'Progress and Promise in Forest Genetics' in *Paper and Science Technology; the Cutting Edge*, Proceedings of the 50th Anniversary Conference of the Institute of Paper Chemistry, Appleton, Wisconsin, pp. 31-60

Durzan, D.J. and Steward, F.C. (1968) 'Cell and Tissue Culture of White Spruce and Jack Pine', *Bimonthly Res. Notes, 24*, 30

Esau, K. (1965) *Plant Anatomy*, John Wiley and Sons, New York

Franclet, A. (1981) 'Rajeunissement et Propagation Végétative des Ligneux', *Annales AFOCEL 1981*, pp. 11-14

Gautheret, R.J. (1934) 'Cultur de Tissu Cambial', *Compt. Rendu. Acad. Sci., 198*, 2195-6

Horgan, K. and Aitken, J. (1981) 'Reliable Plantlet Formation from Embryos and Seedling Shoot Tips of Radiata pine', *Physiologia Plantarum, 53*, 170-5

Hussey, G. (1980) 'Micropropagation', *Gardener, 106*, 286-91

Isikawa, H. (1975) '*In Vitro* Formation of Adventitious Buds and Roots on the Hypocotyl of *Cryptomeria japonica*', *Bot. Mag. Tokyo, 87*, 73-7

Isikawa, H. (1978) 'Effects of Cytokinin and Morphactin on Bud Generation from *Cryptomeria* and *Chamaecyparis* Hypocotyl Segments Cultured *In Vitro*' in T. Akihama and K. Nakajima (eds.), *Long Term Research of Favourable Germ Plasm in Arboreal Crops*, Fruit Research Station, Japan, pp. 142-7

Jansson, E. and Bornman, C.H. (1980) '*In Vitro* Phyllomorphic Regeneration of Buds and Shoots in *Picea abies*', *Phys. Plant., 49*, 105-11

Jansson, E. and Bornman, C.H. (1981) '*In Vitro* Initiation of Adventitious Structures in Relation to the Abscission Zone in Needle Explants of *Picea abies*: Anatomical Considerations', *Phys. Plant., 53*, 191-7

John, A. and Murray, B. (in press)

Kleinschmidt, J. (1974) 'Use of Vegetative Propagation for Plantation Establishment and Genetic Improvement, a Programme for Large Scale Cutting Propagation of Norway Spruce', *N.Z. J. Forest Sci., 4*, 359-66

La Rue, C.D. (1935) 'Cultures of Spermatophyte Tissues', *Am. J. Bot., 22*, 913-17

Loo, S.W. and Wang, F.H. (1943) 'The Culture of Young Conifer Embryos *In Vitro*', *Science, 98*, 544-7

Mapes, M.O., Young, P.M. and Zaerr, J.B. (1981) '*In Vitro* Propagation of Douglas Fir Through the Induction of Precocious Axillary and Adventitious Buds' in the *Abstracts of Proceedings of an International Workshop on In Vitro Cultivation of Forest Tree Species*, Fontainebleau, France

McKeand, S.E. (1981) *Loblolly Pine Tissue Culture – Present and Future Uses in Southern Forestry*, School of Forest Resources, North Carolina State University Technical Report No. 64

Mehra-Palta, A., Smeltzer, R.H. and Mott, R.L. (1977) 'Hormonal Control of Induced Organogenesis from Excised Plant Parts of Loblolly pine (*Pinus taeda* L.)' in *TAPPI Forest Biology Wood Chemistry Conference, 1977*, pp. 15-20

Minocha, S.C. (1980) 'Callus and Adventitious Shoot Formation in Excised Embryos of White Pine (*Pinus strobus*)', *Can. J. Bot., 58*, 366-70

Mott, R.L. (1978) 'Tissue Culture Propagation of Conifers' in *Propagation of Higher Plants Through Tissue Culture*, Proceedings International Symposium, University of Tennessee, April 16-17, 1978, pp. 125-31

Mott, R.L. (1981) 'Trees' in B.V. Conger (ed.), *Cloning Agricultural Plants via In Vitro Techniques*, CRC Press, Florida

Mott, R.L. and Amerson, H. (1981) 'A Tissue Culture Process for the Clonal Production of Loblolly Pine Plantlets', *N. Carolina Agricult. Res. Serv. Tech. Bull. 271*

Mott, R.L., Smeltzer, R.H. and Mehra-Palta, A. (1977) 'An Anatomical and Cytological Perspective on Pine Organogenesis *In Vitro*' in *TAPPI Forest Biology Wood Chemistry Conference, 1977*, pp. 9-14

Patel, K.R. and Berlyn, G.P. (1981) 'Genetic Instability of Multiple Buds of *Pinus coulteri* Regenerated from Tissue Culture', *Can. J. Forest Res., 12*, 93-101

Rediske, J.H. (1978) 'Vegetative Propagation in Forestry' in *Propagation of Higher*

Plants Through Tissue Culture, Proceedings International Symposium, University of Tennessee, April 16-17, 1978, pp. 35-43
Reilly, K. and Brown, C.L. (1976) '*In Vitro* Studies of Bud and Shoot Formation in *Pinus radiata* and *Pseudotsuga menziesii*', *Georgia Forest Res. Paper, 86*, pp. 1-9
Reilly, K. and Washer, J. (1977) 'Vegetative Propagation of Radiata Pine by Tissue Culture. Plantlet Formation from Embryonic Tissue', *N.Z. J. Forest Sci., 7*, 199-206
Reinert, J. (1956) 'Dissociation of Cultures from *Picea glauca* into Small Tissue Fragments and Single Cells', *Science, 123*, 457-8
Schmidt, A. (1924) 'Ueber di Chlorophyllbild im Koniferenembryo', *Bot. Arch., 5*, 260-82
Slankis, V. (1947) 'Influence of Sugar Concentration on the Growth of Isolated Pine Roots', *Nature, 160*, 645-6
Sommer, H.E. (1975) 'Differentiation of Adventitious Buds on Douglas Fir Embryos *In Vitro*', *Proc. Int. Plant Prop. Soc., 25*, 125-7
Sommer, H.E. and Caldas, L.S. (1981) '*In Vitro* Methods Applied to Forest Trees' in T.A. Thorpe (ed.) *Plant Tissue Culture Methods and Applications*, Academic Press, New York, pp. 349-58
Sommer, H.E. and Brown, C.L. (1974) 'Plantlet Formation in Pine Tissue Cultures', *Am. J. Bot. Suppl., 61*, 11
Sommer, H.E., Brown, C.L. and Kormanik, P.P. (1975) 'Differentiation of Plantlets of Longleaf Pine Cultured *In Vitro*', *Bot. Gaz., 136*, 196-200
Sterling, C. (1949) 'Preliminary Attempts in Larch Embryo Culture', *Bot. Gaz., 111*, 90-4
Thomas, M.J., Duhoux, E. and Vazart, J. (1977) '*In Vitro* Organ Initiation in Tissue Cultures of *Biota orientalis* and other species of the Cupressaceae', *Plant Sci. Lett., 8*, 395-400
Tranvan, H. (1979) '*In Vitro* Adventitious Bud Formation on Isolated Seedlings of *Pinus sylvestris*', *Biol. Plant. (Praha) 21*, 230-3
Vazart, J., Conciecao, M. da and Thomas, M.J. (1979) 'Structure Anatomique et Cytologique de l'Hypocotyl du *Biota orientalis* L. au Stade de l'Etalement des Cotylédons', *Rev. Cytol. Biol. Vég., 2*, 83-96
Von Arnold, S. and Eriksson, T. (1979a) 'Bud Induction on Isolated Needles of Norway Spruce (*Picea abies* L. Karst) Grown *In Vitro*', *Plant Sci. Lett., 15*, 363-72
Von Arnold, S. and Eriksson, T. (1979b) 'Induction of Adventitious Buds on Buds of Norway Spruce (*Picea abies*) Grown *In Vitro*', *Physiol. Plant., 45*, 29-34
Von Arnold, S. and Eriksson, T. (1981) '*In Vitro* Studies of Adventitious Shoot Formation in *Pinus contorta*', *Can. J. Bot., 59*, 870-4
Webb, K.J. and Street, H.G. (1977) 'Morphogenesis *In Vitro* of *Pinus* and *Picea*', *Acta Hort., 78*, 259-69
Winton, L.L. and Huhtinen, O. (1976) 'Tissue Culture of Trees' in E.J.P. Miksche (ed.), *Modern Methods in Forest Genetics*, Springer-Verlag, Berlin, pp. 243-64
Winton, L.L. and Verhagen, S.A. (1977a) 'Embryoids in Suspension Cultures of Douglas Fir and Loblolly Pine', *TAPPI Forest Biology and Wood Conference, 1977*, pp. 21-4
Winton, L.L. and Verhagen, S.A. (1977b) 'Shoots from Douglas Fir Cultures', *Can. J. Bot., 55*, 1246-50

3 TISSUE CULTURE OF HARDWOODS

John H. Dodds

The previous chapter has dealt with tissue culture of softwoods (conifers). This chapter will deal with hardwoods, but because of the diversity of this group many examples will be omitted or dealt with in subsequent chapters. As with the gymnosperms the hardwood trees cover an enormous geographical area and a wide range of climatic conditions.

This chapter will attempt to deal with the generalised problems and conditions for growth of hardwoods *in vitro* but it must be kept in mind that some variation will be inevitable with such a broad range of species. The previous chapter reviewed relevant work to date in the context of commercial forestry and its specialised requirements, and a similar brief will be followed in this chapter. Other reviews on different aspects of *in vitro* tree culture can be consulted (Durzan and Campbell, 1974b; Mott, 1978; Winton, 1974a, b; Konar and Nagmani, 1974; Bonga, 1977; Mott and Amerson, 1981; David, 1982).

Initiation and Culture of Callus and Cell Suspensions

Angiosperm tree callus was first obtained by Gautheret (1934, 1937). The callus was not however put into sterile culture and transferred at regular passage intervals. In later studies Gautheret (1948) was able routinely to subculture callus from *Salix*. Further initial callus culture experiments were carried out by Morel (1948), and Jacquiot (1950, 1966). The most common source of material within the plant for the initiation of callus and cell suspension cultures is the cambial zone. This is an area of the plant with a very specific hormonal balance, i.e. high concentration of endogenous cytokinins, and is a tissue that is often already undergoing marked mitotic activity (Savidge, 1982; Savidge and Wareing, 1981). The concentration of hormones is such that callus can often be induced by transferring the explant to a medium containing auxin as the sole hormone source (Nitsch, 1963).

Seasonal changes have a marked effect on the concentration of various plant growth hormones in the cambial zone (Savidge and Wareing, 1981; Sussex and Clutter, 1959; Grasham and Harvey, 1970). It is therefore not surprising to find that cambial explants for initiation of cell and callus cultures are best obtained in the spring months, when the plant material is already prepared for a very rapid increase in the rate of cell division within the cambial zone.

The use of callus cultures and cell suspension cultures *per se* to the application of tissue culture is fairly small; the critical feature is the potential to regenerate a whole new plant from those single cell or callus cultures. The regeneration of plants from callus is known to be dramatically affected by the concentration of plant growth hormones (Philips and Wareing, 1982), and more specifically by the ratio of auxin to cytokinins (Skoog and Miller, 1957). Other compounds do however sometimes have an effect on the morphogenetic aspects of plant regeneration (Murashige, 1974a). Bud formation in callus cultures has been possible, e.g. *Ulmus campestris* (Jacquiot, 1951) and *Citrus* spp. (Altman and Goren, 1971; see also Chapter 7). The formation of callus from apple meristems was carried out by Abbot and Whitley (1974) who later regenerated mature plants from this callus. Trees of *Populus tremuloides* (Winton, 1970) and *Betula pendula* were regenerated from callusing cambial explants.

The preference for vegetative propagation of trees, as opposed to the use of seed, is because generally the genetic characters are better maintained by asexual propagation (Thulin and Faulds, 1968; Murashige, 1974a). However, when a callus stage is involved care must be taken, as genetic aberration can occur (D'amato, 1965, 1978; Thomas, 1981). This instability is not found in multiplying organised cultures (see next section) and so can be avoided. It can however be a useful attribute to the range of cell culture techniques as a way of purposely incorporating genetic variation into a population.

Bud Initiation and Micropropagation

Because of the widespread use of shoot-tip cultures in horticulture, plant pathology and developmental physiology, a misuse of the correct botanical terminology has developed. A variety of terms, such as meristem cultures, meristemming, apical meristem cultures, shoot-tip cultures and stem-tip cultures, have been used; for example orchid growers used shoot tips up to 10 mm in length (Murashige, 1974b). The correct terminology is clearly layed down by Cutter (1965). The apical meristem refers only to the region of the shoot apex lying distal to the youngest leaf primordium, while the shoot apex refers to the apical meristem plus a few subjacent leaf primordia, anything larger than this being a shoot culture.

In the *in vitro* culture of trees the use of meristems or shoot tips derived from buds (or regenerated from callus) has several desirable attributes, as has been mentioned in Chapter 1.

Virus Eradication

If true apical meristems are removed this is a very successful way of removing plant pathogens. This technique is of special importance in those species where the disease is carried by one generation to the next through either seeds or pollen. For trees this technique has been used for 'cleaning up' a number of species.

Meristems have produced virus-free cassava (Kartha and Gamborg, 1975) and *Populus* (Berbee, Berbee and Hildebrandt, 1972). Virus-free *Citrus* has also been produced (Bitters, Murashige, Rangan and Nauer, 1972; see also Chapter 7). In some instances so-called meristem cultures have not removed the pathogen and this is often because the explant is not a true meristem and is too large, thus containing a number of contaminated cells (Smith and Murashige, 1970).

Genetic Resource Storage

To store genetic resources by conventional methods requires a large amount of space and labour (see Chapter 10), but the use of meristem or bud cultures may help in this area (Wilkins, Bengochea and Dodds, 1982; Wilkins and Dodds, 1983). It is possible to store isolated cells of some trees, e.g. *Acer pseudoplatanus* by a cryopreservation technique in liquid nitrogen (Sugawara and Sakai, 1974; Withers, 1980), however, if single cells are stored the problems of genetic aberration in callus result.

Micropropagation

Micropropagation can be described as the *in vitro* multiplication of a plant and it normally involves the hormonal release of dormancy of the axillary buds and their outgrowth. These released axillary buds can then be subcultured into a similar medium and the whole process repeated. It is clear that by using this technique large numbers of plantlets can be produced in a very short period of time. The advantage of a micropropagation system over conventional seed propagation is that it is possible clonally to multiply plants with a desired genotype (Jones *et al.*, 1982). As indicated in the previous chapter, one of the important aspects to be considered in terms of the commercial applications of these techniques to forestry is the cost-benefit of micropropagation. It is difficult to reach an exact figure for the cost of seed production or the cost of micropropagation, but it is clear that *in vitro* methods are likely to cost between 30 and 300 per cent more than conventional methods. The potential gain is also difficult to judge. However, one of the most important features is that clonal propagation allows the multiplication and release of new varieties into the market far quicker than by conventional seed methods. The yield of product (i.e. timber, fruit or oil) may also be significantly higher in cloned material (Jones *et al.*, 1982).

The techniques of micropropagation are well defined in Chapter 6 on fruit tree culture and need not be considered further here.

Rooting of *in vitro* Propagated Material

Transfer of *in vitro* propagated material from the sheltered environment of a culture tube to the soil can be a very traumatic period for the plant. In some cases the transfer has proved to be relatively straightforward with a very high success rate, for example with *Betula pendula* (Hutchinson and Yahyaoglu, 1974) and *Populus tremuloides* (Winton, 1970, 1971). A high humidity environment is required and plants often have to keep the roots above the surface of the medium for functional root hairs to develop (Chalupa, 1974). In a comparative study between solid media and filter paper bridges, rooting was always faster and more prolific on the filter paper bridge technique for a range of *Prunus* and *Pyrus* species tested (Wilkins and Dodds, 1982a).

Vegetative shoots produced by *in vitro* micropropagation techniques normally are devoid of a root system and there are two ways to induce this. The first is to transfer the shoots to a root-induction culture medium and to allow the roots to develop under sterile conditions and then transfer rooted plantlets to a sterile potting compost (Wilkins and Dodds, 1982b). The second method is to take recourse to conventional horticultural methods and regard the *in vitro* shoots as small cuttings. The base of the shoot is dipped into a hormone rooting powder and the plantlets are transferred to a sand bed with mist irrigation. There is no hard and fast rule and it is best to see which method works best with the plant material being studied.

Development of New Technologies with Hardwood Trees

Bearing in mind that the growth of hardwood trees is a major economic programme, it has been shown that increases in yield of only 2-3 per cent could have a great economic impact (Libby, Stettler and Seitz, 1969; Carlisle and Teich, 1971). It is possible that protoplast fusion technology will allow the formation of new hybrids with unrealised economic importance and potential (Winton and Stettler, 1974; Durzan and Campbell, 1974a, b; David and David, 1979; David, 1982; see also Chapter 9).

It is possible that in the longer term trees can be genetically engineered to be more efficient. These new areas need to be actively pursued, but at the same time effort must be given to the basic understanding and technology of regeneration of plantlets from callus, micropropagation and rooting, as these areas must be well defined if genetically manipulated plants are to become clonally propagated on a commercial scale.

References

Abbott, A.J. and Whiteley, E. (1974) '*In vitro* regeneration and multiplication of apple tissues' in *3rd International Conference of Plant Tissue Culture, Abst. 263* (Leicester 1974)

Altman, A. and Goren, R. (1971) 'Promotion of callus formation by ABA in *Citrus* bud cultures', *Plant Physiol., 47*, 844-6

Berbee, F.M., Berbee, J.G. and Hildebrandt, A.C. (1972) 'Induction of callus and trees from stem tip cultures of a hybrid Popular', *In Vitro, 7*, 269-73

Bitters, W.P., Murashige, T., Rangan, T.S. and Nauer, E. (1972) 'Investigations on establishing virus-free *Citrus* plants through tissue culture' in W.C. Price (ed.), *5th International Congress of the International Organ of Citrus Virology, Gainesville, Univ. of Florida*, pp. 267-71

Bonga, J.M. (1977) 'Applications of tissue culture in forestry' in J. Reinert and Y.P.S. Bajaj (eds.), *Plant Cell and Organ Culture*, Springer-Verlag, Berlin, pp. 93-108

Carlisle, A. and Teich, A.H. (1971) 'The costs and benefits of tree improvement programs', *Can. For. Serv. Pub., 1302*

Chalupa, V. (1974) 'Control of root and shoot formation and production of trees from Poplar callus', *Biol. Plant., 16*, 316-20

Cutter, E.G. (1965) 'Recent experimental studies of the shoot apex and shoot morphogenesis', *Bot. Rev., 31*, 7-113

D'amato, F. (1965) 'Endopolyploidy as a factor in plant tissue development' in P. White and A.R. Grove (eds.), *Proceedings of the International Conference on Plant Tissue Culture*, Berkley 1965, pp. 449-62

D'amato, F. (1978) 'Endopolyploidy as a factor in development', *Proceedings of 4th International Plant Tissue Culture Congress, Calgary, Canada 1978*

David, A. (1982) '*In vitro* propagation of gymnosperms' in J.M. Bonga and D.J. Durzan (eds.), *Tissue Culture in Forestry*, Junk, Netherlands, pp. 72-108

David, A. and David, H. (1979) 'Isolation and callus formation from cotyledon protoplasts of pine (*Pinus pineaster*)', *Z. Pflanzenphysiol., 94*, 173-7

Durzan, D.J. and Campbell, R.A. (1974a) 'Prospects for the mass production of

improved stock of forest trees by cell and tissue culture', *Canadian Journal of Forestry Research, 4*, 151-74
Durzan, D.J. and Campbell, R.A. (1974b) 'Prospects for the introduction of traits in forest trees by cell and tissue culture', *N.Z. J. Forestry Sci., 4*, 261-6
Gautheret, R.J. (1934) 'Culture du tissu cambial', *Compte. Rendu. Acad. Sci., 198*, 2195-6
Gautheret, R.J. (1937) 'Nouvelles recherches sur la culture du tissu cambial', *Compte. Rendu. Acad. Sci., 205*, 572-4
Gautheret, R.J. (1948) 'Sur la culture indéfinie des tissus de *Salix caprea*', *Compte. Rendu. Soc. Biol., 142*, 807-8
Grashan, J.L. and Harvey, A.E. (1970) 'Preparative techniques and tissue selection criteria for *in vitro* culture of healthy and rust infected conifer tissues', *U.S. Dept. Agr. Forest Ser. Res. Pap. Int., 82*, 1970
Hutchinson, O. and Yahyaoglu, Z. (1974) 'Das frühe Blühen von aus Kalluskulturen herangezogene Pflänzchen bei der Birko (*Betula pendula*)', *Silvae Genet., 23*, 1-3
Jacquiot, C. (1950) 'Sur la culture *in vitro* de tissue cambial de Châtaignier (*Castanea vesca*)', *Compte. Rendu. Acad. Sci., 231*, 1080-1
Jacquiot, C. (1951) 'Action du meso-inositol et de l'adenine sur la formation de bourgeons par le tissu cambial d'*Ulmus campestris* cultive *in vitro*', *Compte. Rendu. Acad. Sci., 233*, 815-17
Jacquiot, C. (1966) 'Plant tissues and excised organ cultures and their significance in forest research', *J. Inst. Wood Sci., 16*, 22-34
Jones, L.H., Barfield, D., Barret, J., Flook, A., Pollock, K. and Robinson, P. (1982) 'Cytology of oil palm cultures and regenerated plants', *Proceedings 5th International Plant Tissue Culture Congress, Tokyo 1982*
Kartha, K.K. and Gamborg, O.L. (1975) 'Elimination of cassava mosaic disease by meristem culture', *Phytopathology, 65*, 826-8
Konar, R.N. and Nagmani, R. (1974) 'Tissue culture as a method for vegetative propagation of forest trees', *N.Z. J. Forestry Sci., 4*, 279-90
Libby, W.J., Stettler, R.F. and Seitz, F.W. (1969) 'Forest genetics and forest tree breeding', *Ann. Rev. Gen., 3*, 469-94
Morel, G. (1948) 'Recherches sur la culture associée de parasites obligatoires et de tissus végétaux', *Ann. Epiphyties, 14*, 123-234
Mott, R.L. (1978) 'Trees' in B.V. Conger (ed.), *Cloning Agricultural Plants via In vitro techniques*, CRC Press, Florida
Mott, R.L. and Amerson, H. (1981) 'A tissue culture process for the clonal production of Loblolly Pine plantlets', *N. Carolina Agr. Serv. Tech. Bull. No. 271*
Murashige, T. (1974a) Plant propagation through tissue cultures', *Ann. Rev. Plant. Physiol., 25*, 135-65
Murashige, T. (1974b) 'Plant cell and organ culture methods in the establishment of pathogen free stock', *No. 2. A.W. Dimock Lectures,* Dept. of Pathology, *Cornell University, Ithaca N.Y.*
Nitsch, J.P. (1963) 'Naturally-occurring growth substances in relation to plant tissue culture' in P. Maheshwari and N.S. Rangaswamy (eds.), *Plant Tissue and Organ Culture: A symposium*, International Society of Plant Morphologists, Delhi, India, 1963, pp. 144-57
Phillips, D.A. and Wareing, P.F.W. (1982) *The control of growth and differentiation in plants* (3rd edition), Pergamon Press, Oxford
Savidge, R.A. (1982) 'Hormonal factors controlling differentiation', *Histochem. J.* (in press)
Savidge, R.A. and Wareing, P.F.W. (1981) 'Plant growth regulators and the differentiation of vascular elements' in J.R. Barnett (ed.), *Xylem Cell Development*, Castle House, London, pp. 192-236

Skoog, F. and Miller, C.O. (1957) 'Chemical regulation of growth and organ formation in plant tissues cultivated *in vitro*' in *Biological Action of Growth Substances, 11th Symposium of Society for Experimental Biology*, pp. 118-31
Smith, R.H. and Murashige, T. (1970) '*In vitro* development of the isolated shoot apical meristem of angiosperms', *Am. J. Bot., 57*, 562-8
Sugawara, Y. and Sakai, A. (1974) 'Survival of suspension cultured sycamore cells cooled to the temperature of liquid nitrogen', *Plant Physiol., 54*, 722-4
Sussex, I.M. and Clutter, M.E. (1959) 'Seasonal growth periodicity of tissue explants from woody perennial plants *in vitro*', *Science, 129*, 836-7
Thomas, E. (1981) 'Plant regeneration from shoot culture derived protoplasts of tetraploid potato (*Solanum tuberosum*)', *Plant Sci. Lett., 23*, 81-8
Thulin, I.J. and Faulds, T. (1968) 'The use of cuttings in the breeding and afforestation of *Pinus radiata*', *N.Z. J. Forestry, 13*, 66-77
Wilkins, C.P., Bengochea, T. and Dodds, J.H. (1982) 'The use of *in vitro* methods for plant genetic conservation', *Outlook on Agriculture, 11*, 67-73
Wilkins, C.P. and Dodds, J.H. (1982a) 'Effects of various growth regulators on *in vitro* growth of clony shoot tips', *Plant Growth Reg.* (in press)
Wilkins, C.P. and Dodds, J.H. (1982b) 'Effect of physical support on rooting of *Prunus* and *Pynus* shoots *in vitro*', (in preparation)
Wilkins, C.P. and Dodds, J.H. (1983) 'The application of tissue culture techniques to problems of plant genetic conservation', *Science Prog.* (in press)
Winton, L.L. (1970) 'Shoot and tree production from aspen tissue cultures', *Am. J. Bot., 57*, 904-9
Winton, L.L. (1971) 'Tissue culture propagation of European aspen', *Forest Sci., 17*, 48-50
Winton, L.L. (1974a) 'Addendum to the bibliography of tree callus cultures', *General Physiology Notes 19*, Institute of Paper Chemistry, Wisconsin, USA
Winton, L.L. (1974b) 'The use of callus, cell and protoplast cultures for tree improvement', *U.S.-ROC Coop Sci. Prog., Taipei*
Winton, L.L. and Stettler, R.F. (1974) 'Utilization of haploids in tree breeding' in K.J. Kasha (ed.), *Haploids in Higher Plants: Advances and Potential*, Guelph, New York, pp. 259-73
Withers, L.A. (1980) 'Tissue culture storage for genetic conservation', International Board for Plant Genetic Resources/Food and Agriculture Organisation of United Nations, Roma

4 TISSUE CULTURE PROPAGATION OF COCONUT, DATE AND OIL PALM

Jennet Blake

Introduction

The three most important commercial palms are all propagated by seed, and despite continuing selection there is considerable variation between seedlings. The date palm has a limited degree of vegetative propagation by offsets, but there is no natural vegetative means for either coconut or oil palm. However, the realisation over fifteen years ago of the potential for clonal propagation encouraged attempts to obtain cultures of the palms. There was little success at first with date palm (Reuveni, Adata and Lilien-Kipnis, 1972), but later clonal plantlets were obtained by several workers, including Reynolds and Murashige (1979) whose work was successfully followed up by Tisserat (1979, 1981). Progress with oil palm was also slow, but two teams of workers, in England (Jones, 1974) and in France (Rabéchault and Martin, 1976), eventually obtained clonal plantlets. The technique has undergone considerable development (Ahée *et al.*, 1981; Pannetier, Arthuis and Lievoux, 1981), and there has been field evaluation of the clones (Corley, Wooi and Wong, 1981). Eeuwens and Blake at Wye College (1981) have also obtained clonal oil palm plantlets (Figure 4.1). The coconut has not been clonally propagated despite more than a decade of research in several parts of the world.

Figure 4.1: Coconut. Germination of zygotic embryo (a), and in section (b)

Recent reviews which have been concerned with the wider aspects of palm tissue culture include those of Paranjothy (1982) and Reynolds (1982).

Date Palm (*Phoenix dactylifera* L.)

Embryo and seedling tissue of date palm was used by Reuveni *et al.* (1972) in attempts to induce callus for the purposes of propagation, and though a callus was formed with the use of NAA and kinetin, it could not be subcultured successfully. Despite extensive experiments, the problem of browning could not be eliminated. Much later Reuveni (1979) returned to the problem and was able to produce a callus from embryos which eventually produced embryoids.

Several other brief reports suggest that clonal plantlets have been obtained. Smith (1975) used root tips, but the few plantlets obtained were not established. Brochard (1978) obtained callus and also buds directly on roots, but was unable to obtain plantlets from either source. Ammar and Benbadis (1977) developed callus on seedling tissue with the aid of NAA, kinetin and coconut water, and despite problems with browning of the callus, some successful transfers were obtained and plantlets were developed from the callus and established in soil. Details of the transfer medium were not given, but it is suggested that the method was basically similar to that developed for oil palm by Rabéchault and Martin (1976). Ammar and Benbadis (1977) also observed that seedlings treated with IAA and BAP sometimes gave precocious inflorescence development.

Rhiss, Poulain and Beauchesne (1979) were able to obtain multiplication of buds from the tips of offsets and establish them in soil, but did not have any success with inflorescence tissue. Eeuwens (1978) cultured seedling tissue of date palm and was able to initiate callus and obtain roots with NAA.

Callus and embryoids from young embryo tissue were obtained by Reynolds and Murashige (1979) who used a high level of 2,4-D in the presence of activated charcoal. Tisserat (1981) extended this work to a range of tissues from both seedling and mature date palm. He and others investigated the histology of the cultures (Tisserat and Mason, 1980), the cryogenic preservation of the tissues (Tisserat, Ulrich and Finkle, 1981; Ulrich, Finkle and Tisserat, 1982) and published details of his method for the clonal propagation of date palm (Tisserat, 1981). Embryogenic callus was most easily produced from juvenile tissues such as embryos, lateral buds and shoot tips, whereas it was more difficult from stem and inflorescence tissue, which required a longer period of subculturing. Tisserat used a slightly modified MS medium (Murashige and Skoog, 1962) with the addition of activated charcoal at 0.3 per cent, 2,4-D at 100 mg/l and 2iP at 3 mg/l for the initiation and subculture of callus. On this medium the callus became embryogenic and produced bipolar embryoids, which could be multiplied by transfer on the same medium. When transferred to a similar medium without growth substances, the embryoids developed into plantlets which could be established in soil.

At present Tisserat's protocol is the only easily available publication to give full details of a method of vegetative propagation for any of the palms. He has recently given further details of maximising plantlet production by incorporating a low auxin treatment before transfer to a medium devoid of growth substances (Tisserat, 1982).

Oil Palm (*Elaeis guineensis* Jacq.)

The potential of tissue culture propagation of oil palm was first suggested by Staritsky (1970) who excised terminal meristem tissue from a two-year-old palm and obtained some shoot development and initiation of roots, but a complete plantlet was apparently not regenerated.

Embryos cultured by Rabéchault, Ahée and Guénin (1970) produced a callus in the presence of 2,4-D and kinetin over a period of six months, and then produced nodules which tended to multiply and give roots in the presence of IAA. Leaves and embryoids were not formed, though the authors described the nodules as embryoids and suggested that the development of the aerial parts was inhibited by development of the roots, which is a common occurrence in tissue culture.

Culture of tissue from the stem apex and the base of the leaves of 15-month-old palms gave callus in the presence of 2,4-D and kinetin or BAP (Rabéchault, Martin and Cas, 1972) and organogenesis was induced with the aid of 'physiological shocks', i.e. transfers were made from dark to light, the mineral concentration of the medium was altered, the sucrose level was raised, and 2,4-D was replaced by NAA or IAA. Roots and shoots were obtained in the cultures, but it appears that plantlets were not produced.

Callus which could be subcultured was successfully obtained from secondary and tertiary roots of seedlings (Smith and Thomas, 1973) but was remarkably slow-growing. Once the callus was induced, a lower level of auxin was adequate for maintenance. With 2,4-D in the medium the callus was nodular, but with NAA the callus tended to give roots. Ong (1975) cultured the tips of primary roots from seedlings and was able to alter the type of growth obtained by altering the level of growth substances. Thus organised root growth was obtained with a low level of NAA (0.01 mg/l) or by its omission from the medium, an effect observed on both solid and liquid media. In the presence of 2-5 mg/l NAA on solid medium, a slow-growing callus was produced. Although kinetin was added to both media, the level seemed of little importance. Callus induction was increased by growing the root tips to develop a root system before transfer to the higher level of NAA.

In 1974, Jones reported that callus developed from seedling roots (as described by Smith and Thomas, 1973) had produced pearly-white globular structures which elongated and acquired polarity, forming a thick primary root at one pole and a shoot at the other pole. At this stage there was no success with establishment of the plantlets in soil. Jones believed that the globular white structures were similar to the 'embryoids' of Rabéchault *et al.* (1972). 2,4-D was used to induce the callus, but details of the conditions required to produce embryoids were not given, though Jones stated that the observed responses were obtained in conditions which induce embryogenesis in other species. By 1976 (Corley, Barrett and Jones, 1976), clonal plantlets originating from *dura* x *pisifera* seedlings had been established in a 'polybag' nursery in Malaysia. Early field measurements indicated that relative leaf area growth rates and rates of leaf production of clonal plants were similar to those of seedlings grown in the same conditions. All the clonal plants examined appeared to be normal diploids.

Further confirmation of the uniformity within a clone was given by Corley *et al.* (1977, 1979). Three years after planting in the field Corley *et al.* (1981) were able to conclude that a high degree of uniformity existed within clones, and from detailed records of bunch components and oil composition that clones were genetically uniform. They predict that good selected clones should yield 30 per cent more oil per hectare than *dura* x *pisifera* seedlings. Yield data from clones of selected parentage will be available in 1985, when Unilever should make its first commercial sales of clonal plantlets (Corley, 1982).

A popular, but informative, summary of the Unilever work was given by Hawkes (1980).

Whilst the Unilever workers (Jones, 1974; Smith and Thomas, 1973) had been using roots as source tissue for callus, the French workers had been using leaf tissue, with considerable success. In 1976 Martin and

Rabéchault (1976) filed a patent to protect their work, giving considerable details which were published in a condensed form (Rabéchault and Martin, 1976). They used spear leaf tissue from mature palms, which could be obtained without destroying the tree. Each stage of culture was slow and the whole process took up to 2 years, which was probably a similar timescale to that of the Unilever workers, since even in 1980 Hawkes (1980) quoted 18 months.

In the French patent (Martin and Rabéchault, 1976) the leaf tissues were given a quarantine period of 2-4 weeks in the dark on an agitated liquid medium without growth substances. After this they were transferred to an agar medium with 2,4-D at 2-10 mg/l to induce callus, and successive transfers were on a similar medium with the 2,4-D level reduced to 0.2-0.5 mg/l. For differentiation to embryoids the callus was transferred to a medium in which the 2,4-D was replaced by 2.5 mg/l BAP and 0.5 mg/l NAA. Embryoid formation was very variable but was increased in the light. To develop plantlets the embryoids were transferred to a medium lacking growth substances, but if they failed to develop roots, they were given a short treatment with 0.05-0.1 mg/l NAA. Finally, the plantlets were moved to a liquid medium before establishment in pots. Throughout the culture period the basal medium was that of Murashige and Skoog (1962) with the minerals reduced to 40 per cent and the phosphate level doubled. The sucrose level was varied from 1 to 4 per cent and the pH was generally 4.5.

The French workers deliberately extended their research on *in vitro* culture procedures rather than create large amounts of clone material (Kohlenbach, 1977), though a report by Ahée *et al.* (1981) gave a few further details of their culture techniques. Calluses appeared at 60-100 days after isolation of leaf tissue, but the success varied according to the original palm, the part of the leaf used, and also with the hormone concentration of the medium. The proportion of explants bearing calluses was 20-60 per cent. The calluses grew slowly through subculture (an increase of 50 per cent FW in 2 months) and became nodular. Callus stocks have been established from 61 different hybrid trees of *E. guineensis* x *E. guineensis* and *E. guineensis* x *E. melanococca*.

With continued subculture a fast-growing callus (FGC) was formed which had a FW doubling time of 10-20 days or 30-60 days depending on source. The FGC were granular and friable, and had dense areas of meristematic cells amongst larger vacuolated cells and lacunae. FGC can now be established in 6-12 months, and of the 61 trees placed in culture between 1978 and 1980, 22 have now produced FGC.

Embryogenesis of the FGC was obtained by transfer to a medium with both auxin and cytokinin (Hanower and Pannetier, 1982), and by subculture the embryoids were multiplied or grown on into plantlets. The shoot took about 2 months to reach 2 cm, and sometimes needed a rooting treatment. When the shoot had reached 10 cm it was transferred to non-axenic conditions. Embryoids produced in 1976 were planted out in 1978 in the Ivory Coast and in 1980, at the time of first flowering, appeared normal in growth, morphology and karyotype. FGC callus obtained in 1977 has maintained its embryonic potential for 5 years.

By 1981 the ORSTOM-IRHO team were ready to test their laboratory research on an industrial scale in the Ivory Coast (Lioret and Ollagnier, 1981), but it was apparent that the long period from primary explant to clonal plantlet was considered unsatisfactory. In a further report Pannetier *et al.* (1981) described a rapid method in which embryoids appeared within 5-6 months of isolation of the parent tissue, instead of the usual 12-18 months. In a further two months leaves appeared. At present only 4 per cent of primary calluses have produced embryoids in the shortened period, though each callus may produce up to 30 embryoids. This method, which eliminates the FGC stage, has the advantage of maintaining the tissue as callus for the shortest possible time, thus lessening the chance of producing mutants. Further details of the media have been given by Hanower and Pannetier (1982; as shown in Figure 4.2).

Despite the number of reports on the tissue culture of the oil palm, the only detailed protocol for the production of clonal plantlets is that contained in the French patent, in which a number of important details are omitted. More recently, they have published an excellent series of photographs (Pannetier *et al.* 1981), but have omitted details of the medium and culture conditions. The success rate appears to vary, but in their rapid method, embryoids are produced on 4 per cent of 20-60 per cent of the initial cultures, i.e. a true rate of about 1-2 per cent. It is presumably higher in the original method, but even then only about 30 per cent of trees cultured had produced a fast-growing callus by 1980. Unilever have claimed an apparently higher success rate in that cultures have been started from all the palms attempted (Corley *et al.*, 1979). However, as callus lines from only seven mature palms have become embryogenic, the true success rate may be similar to that of the French team.

Despite the apparent low rate of success, both Unilever and IRHO are proceeding with the commercial development of clonal oil palm production (Jones, 1982; Lioret, 1981, 1982). Other laboratories are attempting to develop the technique, but because of the commercial interest it is possible that a satisfactorily detailed protocol may never be published.

Coconut (*Cocos nucifera* L.)

The earliest successful tissue culture experiments with coconut were those of de Guzman (1971) using the embryo of the Makapuno coconut. The embryo of this variety never germinates normally, but when removed from the nut and placed in culture it gave a plantlet which could be grown in the field and was finally shown to yield true-to-type Makapuno nuts (del Rosario and de Guzman, 1982). The embryo culture of de Guzman and her associates gave great encouragement to workers trying to propagate the coconut vegetatively.

Two approaches have been used in attempts to find a method of clonal propagation. One derives from the fact that the inflorescence is the only tissue, apart from the terminal apex, which is meristematic, and that inflorescence tissue in some species may give vegetative plantlets. It has been occasionally observed in nature that the coconut inflorescence became leafy; either the whole inflorescence was converted (Kempanna, 1967) or the individual flower meristems (Davis, 1962), but it was not until 1978 that the successful rooting of such a structure was reported (Sudasrip, Kaat and Davis, 1978).

Immature coconut inflorescence tissue has been successfully cultured (Blake and Eeuwens, 1982) to produce a form of shootlet which could be excised and rooted in culture. However, the 'shootlets' never became truly vegetative and this line of research was not pursued. Other workers have tried to culture inflorescence tissue, but have also been unsuccessful in using it as a means of propagation (de Guzman and del Rosario, 1979; Kuruvinashetty and Iyer, 1979).

The alternative approach for coconut has been to develop callus using a method analagous to that of oil palm. Callus was developed on mature tissues from the terminal meristem (Apavatjrut and Blake, 1977) and from rachis explants (Eeuwens, 1976, 1978), but it could not be subcultured. The Y3 medium, containing both ammonium and nitrate nitrogen at lower levels than in MS medium, and with high levels of potassium and iodine, was developed by Eeuwens and gave good growth of coconut, date and oil palm tissues. A subculturable callus was not obtained until activated charcoal (0.25 per cent) and a high level of 2,4-D (10^{-4} – 10^{-3} mol) was added to the medium. Callus initiation

Figure 4.2: Schematic Representation of Stages in Vegetative Propagation

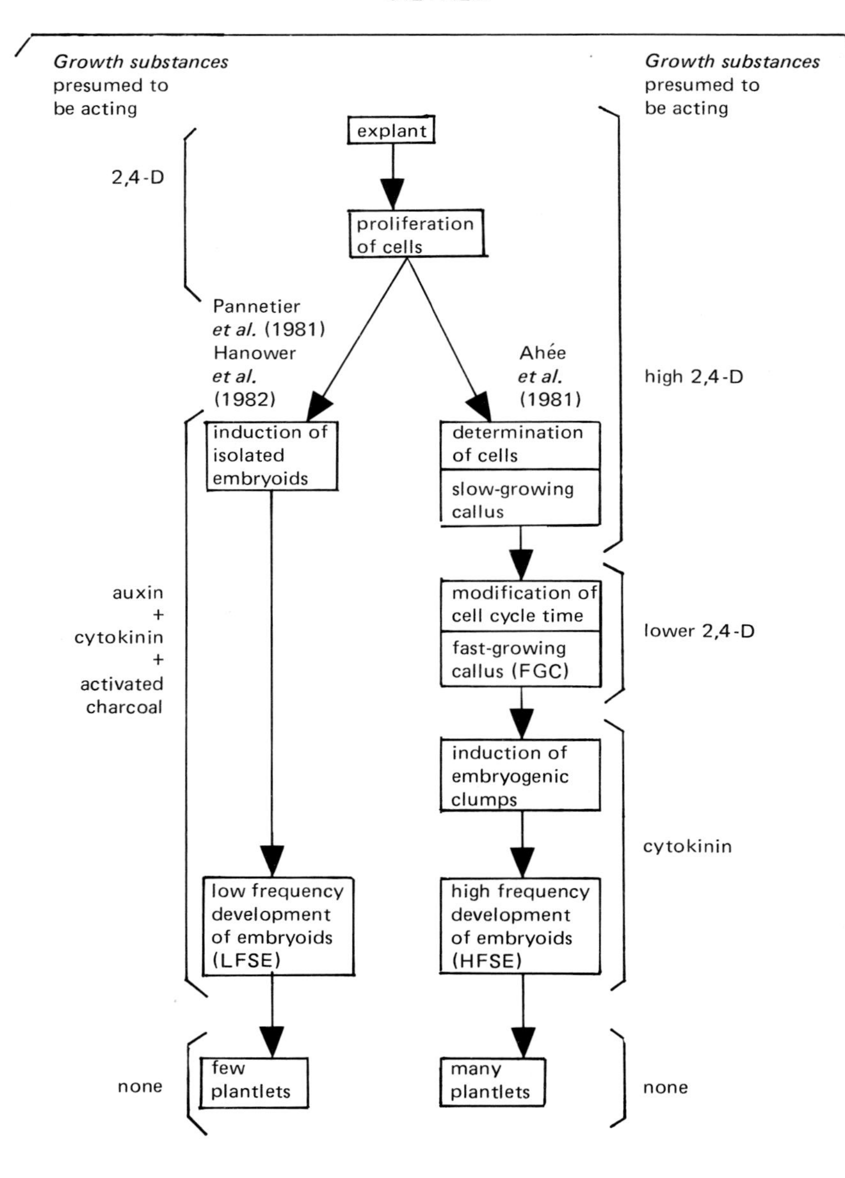

of Oil Palm and Date Palm

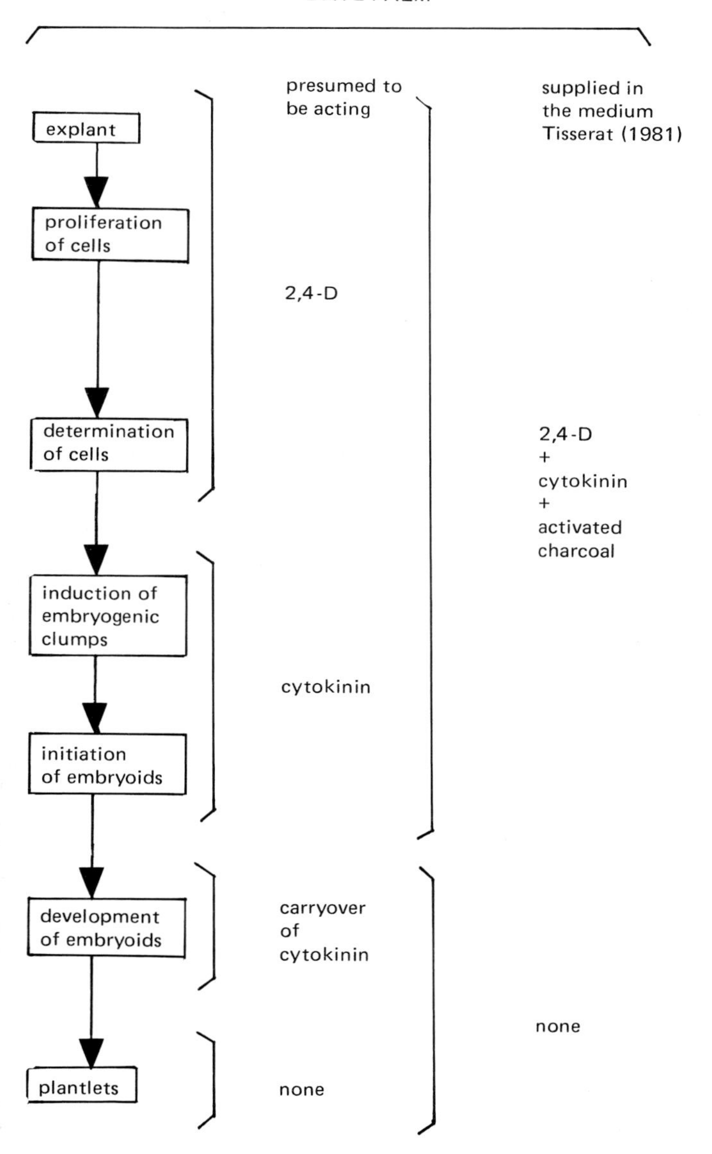

occurred on both a liquid medium and on agar, but it was most satisfactory when the tissue was grown initially in a liquid medium, followed by transfer after 2-4 weeks to an agar medium. Successful callus initiation also occurred when the original explant was finely cut up with a scalpel. The callus was initiated on both juvenile and mature tissue. Subculture of the callus was successful on a similar medium with the same, or a slightly lower concentration of 2,4-D (2.5×10^{-4} – 10^{-4} mol), and was then creamy, firm and nodular. When transferred to a lower level of 2,4-D (approx. 2.5×10^{-6} mol) the callus grew faster, becoming wet and translucent or white and spongy. With repeated subculture on a medium with approx. 10^{-4} mol 2,4-D, the callus became somewhat friable and tended to change structure and become organogenic (Figure 4.4a). It resembled the embryogenic oil palm callus of Ahée *et al.* (1981) and pro-embryogenic tissues appeared to be present. True embryoids which will develop as plantlets have not yet been observed. Similar results have been obtained by Pannetier and Buffard-Morel (1982) also using Eeuwens' Y3 medium.

A similar callus has been observed on embryos by de Guzman, del Rosario and Ubalde (1978) who, in an analogy with orchid culture, suggested that two pathways of plantlet regeneration may be possible: direct plantlet regeneration involving 'protocorm' formation and an indirect regeneration through a callus phase. They appeared to have observed both of these pathways. Noerhadi (1979) and Suryowinoto, Bhiluningputro and Soemario (1979) have also observed a limited amount of callus formation on coconut seedling tissue.

The evidence to be discussed later of Branton (unpublished data) and Pannetier and Buffard-Morel (1983) suggests that embryogenesis in coconut may be obtained by a similar route to that of oil palm.

Critical Stages

A superficially clear pattern exists for the vegetative propagation of date and oil palm, but published details of methods are not always precise, and exact protocol details are often deliberately omitted where commercial ventures are involved. Despite this intentional obscurity, there is a lack of fundamental information on the factors controlling embryogenesis, not only in the palms, but also in other species. A variety of theories have been put forward (see Sharp, Söndahl, Caldas and Maraffa, 1980) but treatments are rarely causative of embryogenesis, which more often seems to be associated with genotype and selection of explant tissue, as in *Zea mays* (Rice, Reid and Gordon, 1979). Although often not specifically discussed, such differences in genotype and source of tissue are undoubtedly important in culturing the palms. Once in culture, there are two critical stages upon which we need to increase our understanding before we can successfully propagate the palms vegetatively: these are callus initiation and induction of embryogenesis.

Callus Initiation

There is relatively little information regarding callus initiation, but the process is generally slow and callus may only be produced on about 10-50 per cent of the explants. Greater success is obtained with seedling tissue than with mature tissue. It seems clear that a high level of auxin is required for callus initiation, but that with date and coconut palm the required level can only be obtained with the addition of activated charcoal. Oil palm tolerates a high level of auxin without charcoal, and the addition of charcoal does not appear consistently to affect either the success rate or the speed of initiation of callus. In all the palms, as in many other monocotyledons, 2,4-D is the most successful auxin, though NAA at a slightly higher level will also produce callus. The disadvantage of NAA for callus induction is that it also acts at a lower level to induce roots, and so there is always a tendency for the callus to produce roots or pneumathodes.

The initial rate of growth of oil palm callus is slow, but after one or two subcultures a faster growing callus is often obtained, though the reason for this is not clear. Since callus induction is slow to start (3 months or more) and the callus is very slow-growing at first, it is possible that the induction medium is sub-optimal. Coconut callus does not appear to have such a clearly marked lag-phase, except perhaps for root callus (Fulford, Passey and Butler, 1979), and generally grows faster than oil palm callus. Date palm is the most rapidly growing palm callus.

Initial treatment of tissue in liquid medium may be advantageous. The French workers have used a liquid pretreatment with shaking for oil palm, whilst Blake and Eeuwens (1982) found with coconut that a 2-4 week period in static liquid medium at a high auxin level before transfer to a solid medium with the same auxin level was more successful in inducing callus than immediate placement on a solid medium. This suggests that the liquid medium allowed better absorption of auxin by the tissue, but as the French treatment was carried out without growth substances it is equally possible that leaching of inhibitory substances or initial reduction of the endogenous auxin level of the tissue were also important. It may be advantageous to place freshly excised tissues in sucrose during sterilisation and rinsing procedures, since they may have been depleted of energy sources for the period between harvesting and preparing for culture.

Smaller explants often have a better survival rate than larger explants, which may be a result of better absorption of nutrients from the medium. Alternatively, a smaller explant may be less affected by correlative effects from other tissues, suggesting that better control in culture may be obtained by the use of even smaller explants in which the original organisation will be more disrupted. Extrapolation to the use of epidermal strips (Tran Thanh Van, 1979) or single cells (Kohlenbach, 1977), from which regeneration has been obtained, suggest this may be the case. Whatever the size of the explant, the ratio of tissue to the volume of the medium and to the air space of the culture vessel may also be important.

Embryogenesis

There are clear assumptions that vegetative propagation of palms proceeds from a callus phase, through a process of somatic embryogenesis, to the production of plantlets. Structures produced by the callus are identified as embryoids in oil palm (Figure 4.3 b, c), date palm and coconut palm. However there are suggestions (Swamy and Krishnamurthy, 1981; Vasil and Vasil, 1980) that embryoids developing in callus cultures, or even directly on somatic tissues, are not equivalent to zygotic embryos, but should be considered as adventitious buds. Since observations in our laboratory with oil palm, and some of the details from other workers (Corley, 1982) suggest that shoots are often formed without roots, and that roots develop adventitiously later, this alternative of organogenesis rather than embryogenesis must be considered.

For practical propagation purposes it does not matter whether the structures produced by callus are embryoids or adventitious shoots. In both cases the tissues have undergone a callus phase and genetic stability cannot necessarily be presumed with either method. If a shoot is produced instead of an embryoid, an extra treatment may, but not necessarily, be required to produce roots, but the end result is a plantlet, often conventionally described as clonal. However, it is not clonal in the strictly horticultural sense of plants propagated by cuttings or tubers, and clonal palm plantlets may show variation from the parent characteristics. Information on this aspect will soon be available from clonal oil palms planted in Malaysia, but nevertheless the technique is already being developed on a commercial scale and preliminary results indicate it will be a successful method of vegetative propagation.

Figure 4.3: Oil Palm. (a) callus initiation on root (left) and leaf (right) explants, (b) organogenesis in subcultured callus with probable embryoids (arrowed), (c) embryoid and/or shoot development, (d) plantlets, multiple (left) single (right)

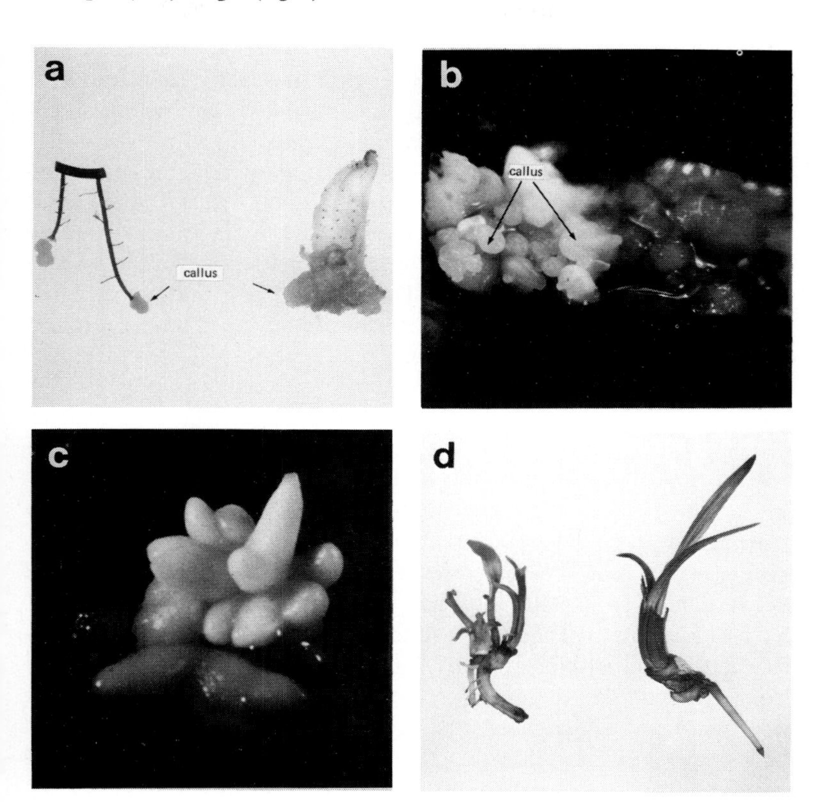

From a fundamental point of view there is a big difference between embryogenesis and the development of a shoot or root meristem by organogenesis. But since fundamental aspects of embryogenesis are so little understood (Thorpe and Biondi, 1981), we are really concerned with the induction of organisation in the callus leading to either (1) the production of a shoot *or* a root meristem (organogenesis), or (2) the development of a somatic embryo with both a shoot *and* a root meristem (embryogenesis). If the latter is not present, as suggested by Swamy and Krishnamurthy (1981), then there is little difference between an embryoid and an adventitious shoot. The following discussion will therefore tend to assume that embryogenesis occurs in the palms, but with the proviso that true somatic embryos may not be formed.

Since the early work of Skoog and Miller (1957) the pattern of auxin-induced callus initiation and cytokinin-induced differentiation has generally been supported, though with interesting exceptions, e.g. alfalfa (Walker, Wendeln and Jaworski, 1979). The palms appear to follow this basic pattern with a high auxin level being used for callus initiation and with cytokinin being used to induce embryogenesis. In some species the primary (callus induction) medium allows the determination of embryogenic cells with the secondary medium permitting embryogenic development (Evans, Sharp and Flick, 1981). For example, in grasses and cereals a high auxin level is used in the primary medium with a lower level of the same auxin, or its elimination, for the secondary medium (Sharp *et al.*, 1980). The assumption is that embryoid or shoot development is determined in the primary medium, and hence the auxin has a dual role of initiating proliferation, followed by inducing embryogenesis, i.e. the determination of the cells to embryogenesis is already present when the initial explant is taken. This is difficult to visualise for the oil palm, where embryogenesis generally occurs after a series of subcultures over many months and without any definite triggering treatment. With the palms, ageing may be important for embryogenesis, whereas other species often lose their embryogenic potential with age (Rice *et al.*, 1979). However the date palm produces embryoids earlier than the oil palm, and may follow the pattern of grasses and cereals more closely. The aim for all the palms must be to find a medium on which the initial explant forms callus as rapidly as possible and allows rapid determination of embryogenic cells, so that embryogenic development can occur when transfer is made to the secondary medium. The view that external application of growth substances can be permissive *or* inhibitory of differentiation, but not determinative (Street, 1978), is a concept which applies well to palm tissue culture.

It is not entirely clear whether cells which are embryogenic actually give rise to embryoids themselves. It seems more probable that the situation is similar to that of carrot in which Jones (1974) concluded that pro-embryogenic (or embryogenic) masses of cells are formed first, and embryoids subsequently develop on these masses. This appears to be the case in date palm, which forms compact aggregates (equivalent to embryogenic tissue) among a matrix of loose friable tissue, and later embryoids are budded from the compact aggregates (Tisserat, 1981). The multiplication of embryoids which is observed in oil palm would then be interpreted as a continued budding of embryoids from the basal embryogenic tissue which persists in the culture, and not as a true multiplication of embryoids. Although it is suggested that a cytokinin is required for embryogenesis in oil palm and date palm it is not clear whether this is really the case since no critical experiments have been carried out. It is difficult to determine whether cytokinins have any role in callus initiation and differentiation in coconut, and therefore the palms may be similar to many other monocotyledons in which cytokinins appear to have no positive effect on shoot formation (Rice *et al.*, 1979).

In coconut, the route for deriving plantlets from callus has not yet been defined though Pannetier and Buffard-Morel have observed structures which they regard as somatic embryoids multiplying by adventive embryogenesis, despite the fact that development to plantlets has not occurred. In our laboratory we have also observed meristematic areas in coconut callus (Figure 4.4 a, b) which have developed into embryoid-like structures (Figure 4.4 c, d, e). Although plantlets have not been formed, the growth pattern of some of these structures has been close to that of a germinating zygotic embryo (Figure 4.4 f compare with Figure 4.1), suggesting that embryoids may have been present. On the other hand, we have observed green leaf-like areas in the callus, and also isolated roots, so that a variety of structures, similar to those shown in cultures of *Atropa belladonna* (Konar, Thomas and Street, 1972), *Dactylis glomerata* (McDaniel, Conger and Graham, 1982) and wheat (Ozias-Akins and Vasil, 1982) may occur when coconut callus becomes organogenic. Many of these structures seen in coconut also occur in oil palm callus where there is often a mixture of normal and abnormal embryoids.

Two possible pathways of embryogenesis have been described for coffee which may be relevant to observations on the palms. Söndahl *et al.* (1977, 1979) have shown that embryoids formed on callus whilst it is on the primary culture medium are few in number: this is designated as LFSE (low frequency somatic embryogenesis); when transferred to the secondary medium, an embryogenic callus is produced which yields large numbers of embryoids: this is HFSE (high frequency somatic embryogenesis). The HFSE route is slower, but produces more embryos. The early work on oil palm appears to describe HFSE, as suggested diagrammatically in Figure 4.2. This HFSE route appears to be the method being developed commercially by Unilever. In contrast, Pannetier *et al.* (1981) describe a rapid method which apparently has a much lower rate of embryoid formation and corresponds with LFSE (Figure 4.2). If this rapid method depends on early development of only a small amount of embryogenic tissue, then it may not have the commercial potential of the slower method which can produce almost unlimited embryoids.

Figure 4.4: Coconut. (a) nodular pro-embryogenic callus, (b) section through callus showing meristematic areas (arrowed), (c), (d), (e) embryoid-like structures (arrowed), (f) haustorium (H) with roots – possible position of shoot (S)

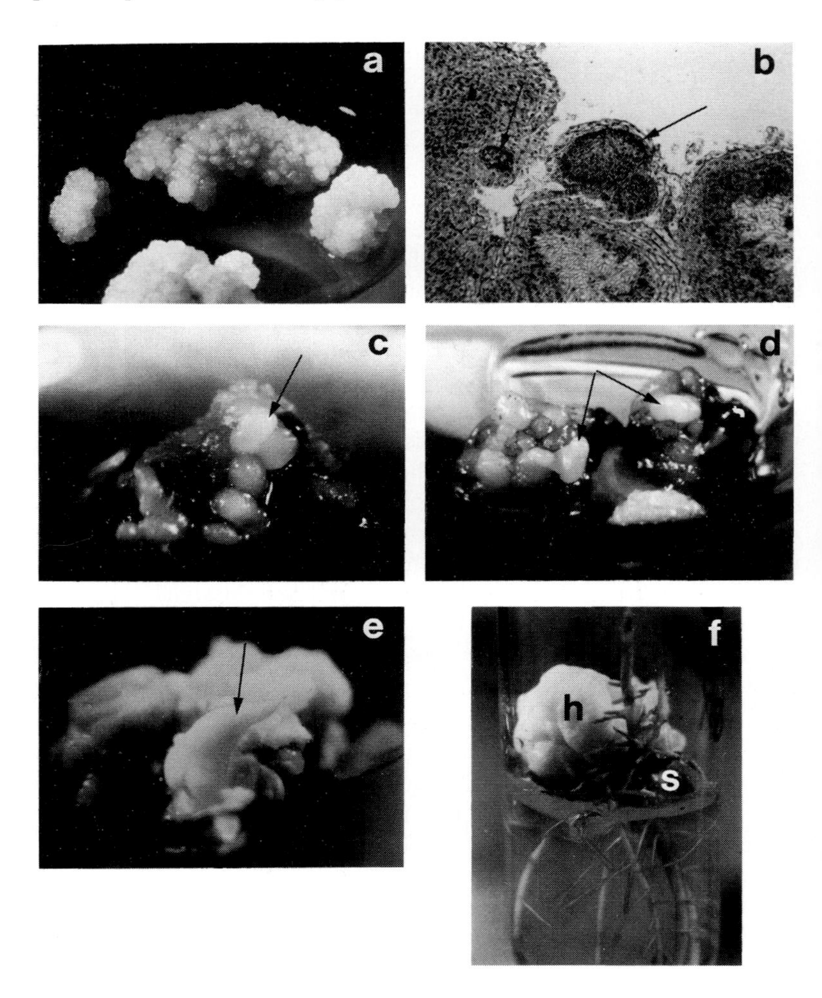

It is clear from the publications on palm tissue culture that embryoid-like structures are formed in culture and lead to the development of plantlets. Since monocotyledonous roots are mostly adventitious in origin, much detailed work would be required to determine if the first root produced is part of a 'true' embryoid or has developed adventitiously at the base of a shoot. Further work on the palms may reveal the full diversity of the organised structures which develop within the callus. Thus organogenesis and embryogenesis may both be occurring in palms, as has already been shown for other monocotyledons (e.g. *Dactylis glomerata* and wheat), and also the dicotyledon, *Atropa belladonna*.

For date and coconut palm it will be many years before it can be shown that all parental characters are transmitted through callus to the offspring, though there are indications of good cytological stability for oil palm (Jones, 1983) and further data should be available by 1985. If the genome remains stable then the callus route for propagation will have two benefits: (1) it will allow almost unlimited propagation of genotypes selected by the plant breeder (Hardon, Corley and Lee, 1982), and (2) it will provide an ideal tissue for cryogenic storage and exchange of germplasm. If the genome fails to remain stable we shall need to culture organised tissue, probably in the form of inflorescence meristems. This method has so far been unsuccessful for propagation (Blake and Eeuwens, 1980), but with further work it could ultimately provide a method of true clonal propagation.

Conclusion

The importance of clonally propagating the palms cannot be overestimated. The current populations have tremendous variability which it may be desirable to maintain in germplasm collections, but for commercial planting it would be advantageous to have reliable high-yielding material of known genotypes. Taking into account the advantages of such material, it may be possible that the cost of tissue culture plantlets will be competitive with seedlings, in which case clonal plantlets may be used for direct planting in commercial plantations. Even were this not so, it is certain that the technique has great advantage for the plant breeder who can multiply desirable parental lines for the production of hybrid seed which may be the most economical method for planting commercial plantations.

As palm seeds are recalcitrant and have a short storage life, tissue culture will be the ideal method of germplasm storage. Here the method already developed is ideal, because a small amount of tissue can be held at the pro-embryogenic or embryogenic stage and then rapidly multiplied as required. International exchange of disease-free germplasm will be greatly simplified by using material in tissue culture.

At the present time the use of tissue culture methods in the development of disease-resistance in palms has been totally unexploited.

The next decade will see important developments in the tissue culture of palms, but because of the long-term nature of the work and the problems still existing with the technique, the full impact is unlikely to be felt until the turn of the century.

Acknowledgements

I am grateful to my colleagues for their help in preparing this review, especially to Drs Richard Branton and Avril Brackpool for stimulating discussions and to Mrs Elizabeth Campling for unfailing patience in searching the literature. I am also grateful to Dr Maro Söndahl for drawing my attention to the similarity between possible embryogenetic pathways in coffee and oil palm.

The first part of this chapter was drafted whilst I was visiting CENARGEN, Brasilia, sponsored by the Interamerican Institute for Cooperation on Agriculture (IICA), and the second part whilst supported by a grant from Deutsche Gesellschaft fur Technische Zusammenarbeit (GTZ) given to Wye College for research on coconut propagation. I am deeply grateful for this support.

References

Ahée, J., Arthuis, P., Cas, G., Duval, Y., Guénin, G., Hanower, J., Hanower, P., Lievoux, D., Lioret, C., Malaurie, B., Pannetier, C., Raillot, D., Varechon, C. and Zuckerman, L. (1981) 'La multiplication végétative *in vitro* du palmier à huile par embryogenèse somatique', *Oléagineux*, *36*(3), 113-18

Ammar, Saïda and Benbadis, Abdelatif (1977) 'Multiplication végétative du Palmier-datier (*Phoenix dactylifera* L.) par la culture de tissus de jeunes plantes issues de semis', *Compte. Rendu. Acad. Sci. Ser. III Vie*, *284*, 1789-92

Apavatjrut, P. and Blake, J. (1977) 'Tissue culture of stem explants of Coconut (*Cocos nucifera* L.)', *Oléagineux*, *32*(6), 267-71

Blake, Jennet and Eeuwens, C.J. (1980) 'Inflorescence tissue as source material for vegetative propagation of the coconut palm', *International Conference on Cocoa and Coconut 1978*, 549-56

Blake, J. and Eeuwens, C.J. (1982) 'Culture of Coconut Palm Tissues with a view to Vegetative Propagation' in A.N. Rao (ed.), *Proceedings COSTED Symposium on Tissue Culture of Economically Important Plants*, Singapore, 1981, pp. 145-8

Brochard, P. (1978) 'Note sur la culture de tissus de Palmier Dattier *Phoenix dactylifera* L.', *Bull. Agronom. Saharienne*, *1*(4), 37-51
Corley, R.H.V. (1982) 'Clonal Plant Material for the Oil Palm Industry', *J. Perak Planters Assoc.*, 1981, 35-49
Corley, R.H.V., Barrett, J.N. and Jones, L.H. (1976) 'Vegetative propagation of oil palm *via* tissue culture', *Malaysian International Agricultural Oil Palm Conference*, pp. 1-8
Corley, R.H.V., Barrett, J.N. and Jones, L.H. (1977) 'Vegetative propagation of oil palm via tissue culture', *Oil Palm News*, *22*, 2-7
Corley, R.H.V., Wong, C.Y., Wooi, K.C. and Jones, L.H. (1981) 'Early results from the first oil palm clone trials', from *The Oil Palm in Agriculture in the Eighties*, 17th-20th June, Kuala Lumpur
Corley, R.H.V., Wooi, K.C. and Wong, C.Y. (1979) 'Progress with vegetative propagation of oil palm', *Planter: Kuala Lumpur*, *55*, 377-80
Davis, T.A. (1962) 'Clonal propagation of the coconut', *World Crops*, Sept./Oct., 253-5
de Guzman, E.V. (1971) 'The growth and development of coconut "Macapuno" embryo *in vitro*. 1. The induction of rooting', *Philippino Agriculture*, *53*, 65-78
de Guzman, E.V. and del Rosario, A.G. (1979) 'Vegetative shoot development and formation in coconut inflorescence tissues cultured in vitro. 1.' FAO Technical Working Party on Coconut Production, Manila, Philippines, 3-8th Dec. 1979
de Guzman, E.V., del Rosario, A.G. and Ubalde, E.M. (1978) 'Proliferative growths and organogenesis in coconut embryo and tissue cultures', *Philipp. J. Coconut Studies*, 1-10
del Rosario, A.G. and de Guzman, E.V. (1982) 'The status of plant tissue culture in the Philippines' in A.N. Rao (ed.), *Proceedings of COSTED Symposium on Tissue Culture of Economically Important Plants*, Singapore, 1981, pp. 293-4
Eeuwens, C.J. (1976) 'Mineral requirements for growth and callus initiation of tissue explants excised from mature coconut palms (*Cocos nucifera*) and cultured *in vitro*', *Physiol. Plant.*, *36*, 23-8
Eeuwens, C.J. (1978) 'Effects of organic nutrients and hormones on growth and development of tissue explants from coconut (*Cocos nucifera*) and Date (*Phoenix dactylifera*) palms cultured *in vitro*', *Physiol. Plant.*, *42*, 173-8
Eeuwens, C.J. and Blake, J. (1981) 'Research on the vegetative propagation of coconut by tissue culture', Report on Ten Years Research under ODA Research Schemes at East Malling Research Station and Wye College (University of London), pp. 17-32
Evans, D.A., Sharp, W.R. and Flick, C.E. (1981) 'Plant regeneration from cell cultures' in J. Janick (ed.), *Horticultural Review Vol. 3*, AVI Publ. Co., Westport, pp. 214-314
Fulford, R.M., Passey, A.J. and Butler, M. (1979) 'Vegetative propagation of coconuts by tissue culture', *Report of East Malling Research Station*, 1979, 184-5
Hanower, J. and Pannetier, C. (1982) '*In vitro* vegetative propagation of the oil palm' in *Proceedings of International Congress of International Association for Plant Tissue Culture*, Tokyo, Japan, 1982
Hardon, J.J., Corley, R.H.V. and Lee, C.H. (in press) 'Breeding and selection for vegetative propagation in the oil palm' in *Symposium on the Improvement of Vegetatively-Propagated Plants*, Long Ashton, Sept. 1982
Hawkes, N. (1980) 'Test tube palms', *Unilever Magazine*, *38*, 43-5
Jones, L.H. (1974) 'Factors influencing embryogenesis in carrot cultures (*Daucus carota* L.)', *Ann. Bot.*, *38*(158), 1077-88
Jones, L.H. (1974) 'Propagation of clonal oil palms by tissue culture', *Oil Palm News*, *17*, 1-8

Jones, L.H. (in press) 'Development of high quality oil palm clones by tissue culture propagation', Royal Irish Academy Seminar, *Plant Tissue Culture in Relation to Biotechnology*, Dublin, Feb. 1982

Jones, L.H., Barfield, D., Barrett, J., Flook, A., Pollock, K. and Robinson, P. (in press) 'Cytology of oil palm cultures and regenerant plants' in *Proc. Int. Congr. Int. Assoc. Plant Tissue Culture*, Tokyo, Japan, 1982

Kempanna, C. (1967) 'A bulbiliferous type of coconut palm', *Mysore J. Agric. Sci.*, *1*, 284-6

Kohlenbach, H.W. (1977) 'Basic aspects of differentiation and plant regeneration from cell and tissue cultures' in W. Barz, E. Reinhard and M.H. Zenk (eds.), *Plant Tissue Culture and its Biotechnological Application*, Springer Verlag, Berlin, pp. 355-66

Konar, R.N., Thomas, E. and Street, H.E. (1972) 'The diversity of morphogenesis in suspension cultures of *Atropa belladonna* L.', *Ann. Bot.*, *36*, 249-58

Kuruvinashetty, M.S. and Iyer, R.D. (1979) '*In vitro* studies for vegetative propagation in coconut', Fifth Session, *FAO Technical Working Party*, Manila, 1979

Lioret, C. (1982) 'Des palmiers éprouvette par millions', *La Recherche*, *13*, 926-8

Lioret, C. and Ollagnier, M. (1981) 'The culture of oil palm tissues *in vitro*', *Oléagineux*, *36*(3), 111-12

Lu, C., Vasil, I.K. and Ozias-Akins, P. (1982) 'Somatic embryogenesis in *Zea mays* L.', *Theor. Appl. Gen.*, *62*(2), 109-12

McDaniel, J.K., Conger, B.V. and Graham, E.T. (1982) 'A histological study of tissue proliferation embryogenesis and organogenesis from tissue cultures of *Dactylis glomerata* L.', *Protoplasma*, *110*(2), 121-8

Martin, J.P. and Rabéchault, H. (1976) 'Procédé de multiplication végétative de végétaux et plants ainsi obtenus', French Patent No. 7628361

Murashige, T. and Skoog, F. (1962) 'A revised medium for rapid growth and bioassays with tobacco tissue cultures', *Physiol. Plant.*, *15*, 473-97

Noerhadi, E. (1979) 'Proliferation of cells on coconut hybrid tissues and the development of embryos grown on a synthetic medium', 5th Session of the FAO Technical Working Party on Coconut Production, Protection and Processing, Manila, Philippines, Dec. 1979

Ong, H.T. (1975) 'Callus formation from roots of the oil palm (*Elaeis guineensis* Jacq.)', *Proceedings of the National Plant Tissue Culture Symposium*, Kuala Lumpur, pp. 25-31

Ozias-Akins, P. and Vasil, I.K. (1982) 'Plant regeneration from cultured immature embryos and inflorescences of *Triticum aestivum* L. (Wheat) – Evidence for somatic embryogenesis', *Protoplasma*, *110*(2), 95-105

Pannetier, C., Arthuis, P. and Lievoux, D. (1981) 'Neo-formation of young *E. guineensis* plantlets from primary calluses obtained on leaf fragments cultured *in vitro*', *Oléagineux*, *36*(3), 119-22

Pannetier, C. and Buffard-Morel, J. (in press) 'First results of somatic embryoid production from leaf tissue of coconut, *Cocos nucifera* L.', *National Coconut Conference*, Malaysia

Pannetier, C. and Buffard-Morel, J. (in press) 'Production of somatic embryos on leaf tissues of coconut, *Cocos nucifera* L.', in *Proceedings of the 5th International Plant Tissue Culture Congress*, Tokyo, Japan, 1982

Paranjothy, K. (1982) 'A review of tissue culture of oil palm and other palms', *PORIM Occasional Paper* No. 3, July 1982 (Palm Oil Research Institute of Malaysia), p. 22

Rabéchault, H., Ahée, J. and Guénin, G. (1970) 'Colonies cellulaire et formes embryoides obtenues *in vitro* à partir de cultures d'embryons de Palmier à Huile (*Elaeis guineensis* Jacq. *var* dura Becc.)', *Compte. Rendu. Acad. Ser.* III *Vie*, *270*, 3067-70

Rabéchault, H. and Martin, J-P. (1976) 'Multiplication végétative du Palmier à huile (*Elaeis guineensis* Jacq.) à l'aide de cultures de tissus foliares', *Compte. Rendu. Acad. Sci. Ser. III Vie*, *283*, 1735-7
Rabéchault, H., Martin, J-P. and Cas, S. (1972) 'Recherches sur la culture des tissus de Palmier à Huile (*Elaeis guineensis* Jacq.)', *Oléagineux*, *27*(11), 531-4
Reuveni, O. (1979) 'Embryogenesis and plantlets growth of date palm (*Phoenix dactylifera* L.) derived from callus tissues', *Plant Physiol.* (suppl.), *63*, 138
Reuveni, O., Adato, Y. and Lilien-Kipnis, H. (1972) 'A study of new and rapid methods for the vegetative propagation of date palms', *49th Annual Date Growers' Conf.*, pp. 17-24
Reynolds, J.F. (1982) 'Vegetative propagation of palm trees' in J.M. Bonga and D.J. Durzan (eds.), *Tissue Culture in Forestry*, Academic Press, London, pp. 182-207
Reynolds, J.F. and Murashige, T. (1979) 'Asexual embryogenesis in callus cultures of palms', *In Vitro*, *15*, 383-7
Rhiss, A., Poulain, C. and Beauchesne, G. (1979) '*In vitro* culture applied to the vegetative propagation of the date palm (*Phoenix dactylifera* L.)', *Fruits d'Outre Mer*, *34*(9), 551-4
Rice, T.B., Reid, R.K. and Gordon, P.N. (1979) 'Morphogenesis in field crops' in K.W. Hughes, R. Henke, M. Constantin (eds.), *Propagation of Higher Plants through Tissue Culture*, Proceedings of the University of Tennessee Symposium, 1978, USDOE Pub. (CONF-7804111), pp. 262-77
Sharp, W.R., Söndahl, M.R., Caldas, L.S. and Maraffa, S.B. (1980) 'The physiology of *in vitro* asexual embryogenesis', *Hort. Rev.*, *2*, 268-310
Skoog, F. and Miller, C.O. (1957) 'Chemical regulation of growth and organ formation in plant tissues cultured *in vitro*', *Symposia for the Society of Experimental Biology*, *11*, 118-30
Smith, S.N. (1975) 'Vegetative propagation of the date palm by root tip culture', *Bull. Agronom. Saharienne*, *1*(3), 67
Smith, W.K. and Thomas, J.A. (1973) 'The isolation and *in vitro* cultivation of cells of *Elaeis guineensis*', *Oléagineux*, *28*(3), 123-7
Söndahl, M.R. and Sharp, W.R. (1977) 'High frequency induction of somatic embryos in cultured leaf explants of *Coffea arabica* L.', *Z. Pflanzenphysiol.*, *81*(5), 395-440
Söndahl, M.R., Spahlinger, D.A. and Sharp, W.R. (1979) 'A histological study of high frequency and low frequency induction of somatic embryos in cultured leaf explants of *Coffea arabica* L.', *Z. Pflanzenphysiol.*, *94*(2), 101
Street, H.E. (1978) 'Differentiation in cell and tissue cultures – regulation at the molecular level' in H.R. Schutte and D. Gross (eds.), *Regulation of Developmental Processes in Plants*, Blackwell, London, pp. 192-218
Staritsky, G. (1970) 'Tissue culture of the oil palm (*Elaeis guineensis* Jacq.) as a tool for its vegetative propagation', *Euphytica*, *19*, 288-92
Sudasrip, H., Kaat, H. and Davis, A. (1978) 'Clonal propagation of the coconut *via* the bulbils', *Philipp. J. Coconut Studies*, *III*(3), 5-14
Suryowinoto, M., Bhiluningputro, W. and Soemario, L. (1979) 'Some coconut tissue culture experiments', *5th Session of the FAO Technical Working Party on Coconut Production, Protection and Processing*. Manila, Philippines, Dec. 1979
Swamy, B.G.L. and Krishnamurthy, K.V. (1981) 'On embryos and embryoids', *Proceedings Indian Academy Science Plant Science*, *90*(5), 401-12
Thorpe, T.A. and Biondi, S. (1981) 'Regulation of plant organogenesis' in K. Maramorosch (ed.), *Advances in Cell Culture*, Vol. 1, Academic Press, New York, pp. 213-39
Tisserat, B. (1979) 'Propagation of Date Palm (*Phoenix dactylifera* L.) *in vitro*',

J. Exp. Bot., *90*(119), 1275-83

Tisserat, B. (1981) 'Date palm tissue culture', U.S.D.A. Agricultural Research Service, Science and Education Administration, AAT-W-17. August 1981, p. 50

Tisserat, B. (1982) 'Factors involved in the production of plantlets from date palm callus cultures', *Euphytica*, *31*, 201-14

Tisserat, B. and de Mason, D.A. (1980) 'A histological study of development of adventive embryos in organ cultures of *Phoenix dactylifera* L.', *Ann. Bot.*, *46*, 465-72

Tisserat, B., Ulrich, J.M. and Finkle, B.J. (1981) 'Cryogenic preservation and regeneration of date palm tissue', *Hort. Sci.*, *16*(1), 47-8

Tran Thanh Van, K.T. (1981) 'Control of morphogenesis in *in vitro* cultures', *Ann. Rev. Plant Physiol.*, *32*, 291-311

Ulrich, J.M., Finkle, B.J. and Tisserat, B.H. (1982) 'Effects of cryogenic treatment on plantlet production from frozen and unfrozen date palm callus', *Plant Physiol.*, *69*(3), 624-7

Vasil, I.K. and Vasil, V. (1980) 'Clonal propagation' in I.K. Vasil (ed.), *Perspectives in Plant Cell and Tissue Culture*, Pt. A, Academic Press, New York, pp. 145-73

Walker, K.A., Wendeln, M.L. and Jaworski, E.E. (1979) 'Organogenesis in callus tissue of *Medicago sativa*. The temporal separation of induction processes from differentiation processes', *Plant Sci. Lett.*, *16*(1), 23-30

5 CULTURE OF ORNAMENTAL TREES

Teresa Bengochea

A problem of nomenclature exists over this particular area of tree tissue culture. It is hard to come up with a hard-and-fast definition of an ornamental. There are a number of trees that may have economic importance in their own right as fruit trees or as timber crops but that are still used to provide pleasant surroundings; this creates a problem of definition as to what is strictly an ornamental tree. The simplest definition of an ornamental tree is that it is a tree grown purely for aesthetic value and not being grown for the harvest of fruit or wood for commercial gain. This chapter is deliberately brief, as other relevant material is covered in other chapters of this book.

The potential market for ornamental trees is an extremely large one as it covers local and national governments (parks, road islands, etc.), businesses and the home garden. It follows therefore that any move to increase the propagation rate of these materials would have very important economic implications (Durzan and Campbell, 1974; Winton, 1972, 1974). At the present time a large amount of the material sold is propagated either by seed or by conventional horticultural cuttings. There are a number of drawbacks to these methods.

Number of Plants to be Produced

If an extremely large number of plants are required in a relatively short period of time then obviously the most effective method for producing that material is seed. There is however a major problem that often the seed for growth of ornamental trees is either recalcitrant (loses its viability very quickly) or is highly heterozygous genetically. This can mean that there is an enormous variation in the standard of the material produced. To avoid the problems associated with seed, ornamental trees are sometimes propagated from woody cuttings by conventional horticultural techniques. The great difficulty with this type of propagation is that it is both slow and labour intensive. The number of plants that can be produced from cuttings is relatively small and to 'clone up' a new variety for release onto the market could take many years.

Table 5.1: List of Woody Ornamental Plants in Culture at Long Ashton Research Station (D.R. Constantine 21.5.82)

	CE	SM		R	PE
Acer platanoides 'Royal Red'	+	0		0	0
A. saccharinum 'Pyramidale'	+	0		0	0
Arundinaria viridistriata	+	0		0	0
Berberis darwinii	+	+	B5 glu 3.0 BAP	0	0
Camelia sinensis	+	+	B5 glu 3.0 BAP	+	–
Chamaepericlymenum canadense (= *Cornus canadensis*)	+	+	MS glu 0.5 BAP	+	+
Cotinus coggygria 'Royal Purple'	+	+	B5 glu 2.0 BAP	0	0
Daphne × *burkwoodii*	+	+	MS suc 0.05 BAP	+	+
D. cneorum	+	+	B5 suc 0.05 BAP	+	0
D. laureola	+	+	MS suc 0.1 BAP	+	0
D. mezereum	+	+	MS suc 0.1 BAP	0	0
D. odora 'Aureomarginata'	+	+	MS suc 0.1 BAP	+	+
D. pontica	+	+	MS suc 0.1 BAP	0	0
D. retusa	+	+	MS suc 0.1 BAP	+	+
D. tangutica	+	+	MS suc 0.1 BAP	+	+
Disanthus cercidifolius	+	0		0	0
Distylium racemosum	+	+	B5 glu 3.0 BAP	0	0
Embothrium coccineum lanceolatum	+	+	MS suc 2.0 BAP	+	+
Garrya elliptica	+	+	MS suc 2.0 BAP	+	–
Hamamelis virginiana	+	+	B5 glu 3.0 BAP	0	0
Ilex aquifolium	+	+	MS suc 3.0 BAP	+	+
I. a. 'J.C. van Tol'	+	+	MS suc 3.0 BAP	+	+
Lapageria rosea	+	+	MS suc 3.0 BAP	–	–
Liquidambar styraciflua	+	+	MS suc 2.0 BAP	+	+

Magnolia × soulangiana	+	+	B5 glu 3.0 BAP	+	+
Mutisia oligodon	+	+	B5 glu 0.3 BAP	0	0
Ruta graveolens 'Jackmans Blue'	+	+	MS suc 0.05 BAP	+	0
Schizophragma hydrangeoides	+	+	MS suc 3.0 BAP	+	+
Skimmia japonica	+	+	MS suc 2.0 BAP	+	+
Vi burnum × juddii	+	+	MS suc 3.0 BAP	+	+

Key:
CE = culture establishment
SM = shoot multiplication
MS = Murashige Skoog
B5 = Gamborg B5
suc = sucrose
glu = glucose
0.05 = 0.05 μ mol
BAP = BAP
R = rooting
PE = plant establishment
+ = positive response
– = negative response
0 = not tested

By isolating meristems and micropropagating (see Chapter 3) it is possible to achieve multiplication rates running up into millions per year from a single plant (Wilkins and Dodds, 1982). This ability to clone plants so quickly reduces the unit cost very significantly and also means that new varieties can be brought onto the market very quickly. Material that is cloned in this way is genetically uniform and is therefore much easier to handle on a large scale than seed-derived material.

In view of the fact that these micropropagated cultures are often started from meristems there is an added advantage that the material is 'clean'. The eradication of virus by meristem cultures has been described (Smith and Murashige, 1970) and the material, by nature of the culture technique, is free from bacterial and fungal infection.

Methods

The choice of culture medium for micropropagation of ornamentals is somewhat empirical. The two most popular choices are Murashige and Skoog (1962) salts or Gamborg (1966) B5 medium. Shoot-tips or true meristems are dissected out and transferred to the surface of the medium for either micropropagation or rooting. The critical feature of the media is the type and concentration of hormone that is added and some details of this are given in Table 5.1 (Abbot *et al.*, 1982).

Material

Table 5.1 gives a comprehensive list of the material that is currently under study at Long Ashton Research Station, near Bristol, England. The list is fairly extensive but really only touches a fraction of the material that could be available for culture.

Potential for the Future

Large Scale Propagation

For many species it may become commercially advantageous to produce the material by micropropagation; by these methods the material will be genetically more uniform and will also be clean and healthy. The ability to produce large amounts of genetically uniform material in a very short period of time will allow the faster release of newly bred ornamental tree varieties onto the market.

Selection of New Material

The use of single cell technology may make it possible to select genotype of ornamentals suited to adverse environments, i.e. disused mine tips, etc. It may also be possible to use single cell techniques (Chapter 9) to select new varieties of ornamentals. One of the great advantages of an ornamental is that the selection criterion is a simple one, i.e. whether it looks good.

The technique of tissue culture can therefore be applied to ornamental trees both to increase the number of those we already have and to develop new ones.

References

Abbott, J., Constantine, D., Belcher, A. and Wiltshire, S. (1981) 'Micropropagation of woody ornamentals', Long Ashton Research Station, Bristol.

Dodds, J.H. and Roberts, L.W. (1982) *Experiments in Plant Tissue Culture*, University Press, Cambridge, U.K.

Durzan, D.J. and Campbell, R.A. (1974) 'Prospects for the mass production of improved stock of forest trees by cell and tissue culture', *Can. J. Forestry Res.*, *4*, 151-74

Gamborg, O.L. (1966) 'Aromatic metabolism in Plants', *Can. J. Biochem.*, *44*, 791-9

Murashige, T. and Skoog, F. (1962) 'A revised medium for rapid growth and bioassays with tobacco', *Physiol. Plant.*, *15*, 473-97

Smith, R.H. and Murashige, T. (1970) 'In vitro development of the isolated shoot apical meristems of angiosperms', *Am. J. Bot.*, *57*, 562-8

Wilkins, C.P. and Dodds, J.H. (1982) 'The application of tissue culture techniques to plant genetic conservation', *Scientific Prog.* (Oxf), *68*, 281-307

Winton, L.L. (1972a) 'Annotated bibliography of somatic conifer callus cultures', *General Physiology Notes*, Institute of Paper Chemistry, *16*, 19-20

Winton, L.L. (1972b) 'Bibliography of somatic callus cultures from deciduous trees', *General Physiology Notes*, Institute of Paper Chemistry, *16*, 21-25

Winton, L.L. (1974) 'The use of callus, cell and protoplast cultures for tree improvement', *US-ROC Co-operative Science Program*, Taipei

6 TISSUE CULTURE PROPAGATION OF TEMPERATE FRUIT TREES

Christopher P. Wilkins and John H. Dodds

Since very early times, man has cultivated those species of fruit and nut trees indigenous to the temperate zone latitudes. Archaeological excavations have indicated that the apple was known to man in the Stone Age (Zagaja, 1970), whilst written records exist from 1100 BC detailing the culture of certain oriental pear species (Westwood, 1978). The ancient Greeks and Romans described several varieties of apples, plums, grapes and cherries, and were conversant with methods of propagation by both budding and grafting (Sekowski, 1956; Zielinski, 1955). Similarly, peaches and apricots were domesticated first in China where their cultivation was known from at least 2000 BC (Sekowski, 1956).

Most fruits of the temperate zone originated in the Northern Hemisphere of the Old World, whilst only a few are native to either North or South America and none to Australia (Vavilov, 1926; Zielinski, 1955). As a result of man's early migrations, fruits were carried with him and subsequently established in newly occupied areas, and in this way spread rapidly over large areas of the Old World very early in human history (Zagaja, 1970). Subsequent travels by man later served to establish deciduous fruit and nut tree species in all regions of the world wherever climate, soil and moisture conditions were suitable. For example, the European pear is adapted to such diverse areas as Australia, South Africa and the Western United States (Westwood, 1978). The limits of distribution of any particular deciduous fruit tree species are determined primarily by climatic factors. Westwood (1978) has listed the general climatic requirements for temperate zone trees and shrubs as follows:

(1) Winter temperatures must not be so cold that they kill the plants.
(2) Winters must be cold enough to give buds adequate chilling to break winter rest.
(3) The growing season (number of frost-free days) must be long enough to mature the crop.
(4) Temperature and light during the growing season must be adequate for the species in question to develop fruit of good quality.

In general, these climatic restraints mean that temperate fruit trees are confined mainly to the middle latitudes between about 30° and 50°. However, cultivation may extend to lower latitudes at high elevations and to higher latitudes in regions where large bodies of water have a moderating influence. A comprehensive discussion of all aspects of temperate zone pomology has been given by Westwood (1978) and references cited therein.

From the pomological point of view, temperate fruits may be classified into four major groups:

(1) Pome fruits: apple (*Malus* spp.); pear (*Pyrus* spp.) and quince (*Cydonia* spp.) together with certain other species and genera which are of possible use as rootstocks, such as Chinese quince (*Chaenomeles*); medlar (*Mespilus*); hawthorn (*Crataegus*); mountain ash (*Sorbus*) and service berry (*Amelanchier*).
(2) Stone fruits: peach, nectarine, plum, cherry, apricot and almond (*Prunus* spp.), and many species used only as rootstocks or ornamentals.
(3) Tree nuts: walnut (*Juglans* spp.), hickory (*Carya* spp.), pecan (*Carya illinoensis*), hazel (*Corylus* spp.), pistachio (*Pistacia vera* L.) and chestnut (*Castanea* spp.).
(4) Berry fruits: gooseberry (*Ribes grossularia* L.), grape (*Vitis* spp.), currants (*Ribes* spp.), raspberries and blackberries (*Rubus* spp. [Tourn.] L.) and cranberry (*Vaccinium* spp.).

Other important deciduous fruits which have been classed as belonging to the category labelled 'temperate' (Westwood, 1978), are mulberries and figs, both members of the Family *Moraceae*; and also fruits of minor economic importance such as persimmon (*Diospyros* spp.), Northern pawpaw (*Asimina triloba* [L.] Dun.), pomegranate (*Punica granatum* L.) and jujube (*Zizyphus jujuba* Mill.).

As will be evident from the above brief introduction, the temperate zone tree fruits are an extremely heterogeneous group of plants with respect to their origin, taxonomy, ecological requirements and breeding system (Zagaja, 1970). Furthermore, since the deciduous tree fruits do not produce true-to-type from seed, and are therefore highly heterozygous, they must be propagated as clones. Vegetative propagation is therefore an absolute requirement for all commercially important fruit tree cultivars.

The major fruit species of the world are now nearly all asexually propagated by the processes of budding and grafting, in which a single bud or short piece of shoot (the scion) is cut from the desired variety of tree that is to be increased, and is then applied to an already established tree (the rootstock) so as to form a composite tree. This roundabout method of propagation is necessary since hardwood cuttings of most fruit trees are extremely difficult to root. The use of such techniques is therefore the only feasible way to obtain a number of clones of a desirable fruit tree cultivar.

Prior to the last half century, trees to be used as rootstocks were raised indiscriminately from any seedlings of the same species which happened to be available (Hudson, 1982). Since then, however, an enormous amount of research effort has been expended into the development of rootstocks designed for a specific function. Such research has resulted in there being available a multitude of rootstock types, the choice of which can have a profound effect on the precocity and productivity of the major fruit species: affecting factors such as growth control, tolerance to soil and climatic variables, resistance to soil pests and pathogens, yield efficiency, anchorage and ease of propagation. Perhaps the best known of the rootstock types are the M and MM series of apple rootstocks. The M series (M1, M2, M3, etc.) of rootstocks were developed at the East Malling Research Station, England and comprise both dwarfing (M27, M9, M26, M7) and invigorating (M25, M1) stocks, as well as intermediate types. The MM series (MM101, MM102, etc.) were developed jointly by the East Malling and Merton Stations, and were specifically developed for resistance to woolly aphid (*Eriosoma lanigerum*) (Tukey, 1964). In the case of pears, the choice of rootstock can greatly influence fruit quality, the use of an incompatible rootstock often giving rise to specific fruit disorders such as 'cork spot', which lowers the grade of the fruit. A full discussion of the function, propagation and performance of different rootstocks has been given by Westwood (1978).

Although conferring enormous benefits to fruit growers, the use of such sophisticated propagation techniques also has several inherent disadvantages. One of the most serious of these is the fact that fruit tree viruses have become widely disseminated through the use of such techniques. As discussed by Hudson (1982), such viruses can have many effects on the size, colour and quality of fruits, and as such are fairly easy to recognise. However, other viruses are more insidious in their effects, and whilst not showing obvious visual symptoms, may affect the growth, vigour, productivity and longevity of trees.

This is especially serious for fruit-tree growers since the establishment of a fruit-tree orchard involves a high capital expenditure and is expected to give financial returns over many years of fruit production.

In an attempt to halt the further spread of virus-infected material, an extensive programme of work by the Long Ashton and East Malling Research Stations (two of the Agricultural Research Council's institutes in the U.K.) has resulted in the development of the 'EMLA' scheme for fruit-tree stocks and scions. Material designated as being of 'EMLA' quality is certified as being virus-free and therefore allows the production of first-class virus-free fruit trees.

However, all such highly selected stocks must of necessity be propagated by asexual means; usually by the techniques of stooling or layering. This process takes three years, and demands expensive nursery facilities and skills (Jones, Pontikis and Hopgood, 1979).

It will be evident from the preceding introduction, that those methods of fruit tree propagation currently in use at the present time have many limitations and shortcomings. The next section will discuss ways in which techniques of *in vitro* propagation could possibly be utilised to overcome some of these problems, and will review possible future applications of such techniques.

Applications of in vitro **techniques**

Several workers have briefly listed the possible advantages to be gained from the *in vitro* culture of woody fruit plants (Pierik, 1975; Abbot and Whiteley, 1976; Zimmerman, 1978; Lane, 1978, 1979; Jones, 1979). Most of these applications are a direct result of the extremely rapid rates of multiplication made feasible by tissue-culture systems.

Due to the fact that the conventional methods of asexual propagation of fruit trees are slow and demand specialist nursery facilities, many newly released cultivars from breeding programmes are often in very short supply. This applies both to rootstock and scion cultivars. Furthermore, dwarf cultivars have very slow-growing shoots and thus could not be expected to produce enough shoot cuttings for rapid propagation. The facility for rapid clonal propagation provided by *in vitro* techniques would therefore serve to hasten the availability of new selections from breeding programmes. Similarly, the rapid clonal propagation of self-rooted virus-free scion varieties would obviate the need for rootstocks, and thereby give significant economic benefits in modern high density orchard systems. Such techniques may also make the 'meadow orchard' system proposed by Hudson (1971), into an economically viable concept. (This is a method of fruit production whereby trees are grown very close together (up to 75 000 per hectare).) Since it would be uneconomical to plant budded trees at such a density, the use of self-rooted scion cultivars is therefore the only alternative. A single stem from each plant is induced to form flower buds much earlier than normal, and the resulting fruit crop is then mechanically harvested (Hudson, 1982). The major advantage (in financial terms) of meadow orchardry is that all operations may be performed by machine. Although such systems are still speculative as far as apple growing is concerned, very heavy yields have already been produced under experimental conditions, and the technique has already been introduced on a commercial scale for growing peaches. Although much progress in this field has already been made using new *in vivo* techniques for the rooting of apple scion cuttings, it is likely that planting on such a massive scale will not become economically feasible until the development of reliable micropropagation techniques.

Another prospective application of *in vitro* techniques is the rapid clonal multiplication of desirable but scarce rootstocks which have been certified as virus-free, such as those produced by the 'EMLA' scheme. The use of specialist meristem culture techniques should also facilitate the initial production of virus-free plants, since the fact that the meristematic part of a shoot apex remains virus-free has been known for many years (Morel and Martin, 1955).

Other areas where tissue culture techniques may prove to have considerable benefits, are in the fields of conservation of plant genetic resources via *in vitro* germplasm storage, and germplasm exchange. Such *in vitro* techniques would facilitate the preservation of special breeding lines, or of novel types such as may be produced in mutation-breeding experiments. Alternatively, *in vitro* techniques would allow the rapid clonal multiplication of such types for prompt evaluation in field trials.

The subject of genetic conservation is an extremely active field of research at the present time, and the application of *in vitro* techniques for the preservation of threatened fruit-tree germplasm resources is discussed elsewhere in this volume. Related to the topic of germplasm storage, is the concept of the international exchange of disease-free plants or of valuable genetic stocks as *in vitro* cultures. The aseptic nature of such cultures would overcome the quarantine regulations of many countries which prohibit both the import and export of rooted material.

Other possible uses of tissue culture techniques include more specialised, and at present mostly speculative, applications, such as the production and maintenance of haploids for use in breeding programmes; or the use of techniques involving the isolation and culture of protoplasts and the subsequent production of somatic hybrids. The development of reliable methods of protoplast isolation and culture would make feasible the use of biotechnological techniques involving the actual manipulation of genetic material.

Development of Micropropagation Methods

In the years which have elapsed since Haberlandt's hypothesis of the totipotency of plant cells was finally realised, there has been an enormous upsurge of interest in the possible uses and applications of *in vitro* techniques. The field of plant cell, tissue and organ culture has now grown to be of immense size, scope and importance, and this volume is a testimony to the specialised nature of the present day applications of plant tissue culture techniques. It was inevitable that early investigators should turn their attentions to the problems of the *in vitro* propagation of woody plants, and several early workers (Geissbühler and Skoog, 1957; Haissig, 1965) stressed the prospective potential of such an application. However, in spite of a great deal of early work concerned with the *in vitro* culture of various woody species, it is only comparatively recently that reliable techniques have been developed that will make possible the rapid clonal propagation of the economically important species of temperate fruit trees on a commercial scale.

Early workers in the field were concerned primarily with the use of meristem culture techniques to eliminate systemic virus infections from species such as gooseberry (Jones and Vine, 1968) and apple (Walkey, 1972). Other workers centred their investigations on the effects of cytokinin compounds on apple shoot-tips (Jones, 1967; Pieriazek, 1968). Several later workers showed that it was possible to culture shoot-tips of apple *in vitro*, but there was no multiplication of shoots, and few rooted (Dutcher and Powell, 1972; Elliot, 1972; Quoirin, 1974). A later report (Abbot and Whiteley, 1976) showed that shoots of apple seedlings and of the scion cultivar Cox's Orange Pippin could be induced to multiply *in vitro* by about 10-fold per month. However, shoots were small, difficult to root, and plants were not established *in vivo*.

It was not until well into the last decade that Jones, Hopgood and O'Farrell (1977) demonstrated that virus-free shoots of the apple rootstock M26 could be rapidly multiplied and rooted *in vitro*, then subsequently established under non-sterile conditions with a high degree of success. Such work indicated that more than 60 000 plantlets could be produced from a single shoot-tip over an 8-month culture period. At about the same time, workers such as Boxus and Quoirin (1977), turned their attention to the problems concerned with the *in vitro* propagation of various *Prunus* species. However, the rooting percentages they achieved were low (30 per cent), and most of the resulting plantlets died on transferring to pots of soil. Subsequent work by Jones and Hopgood (1979), applied the techniques developed from the *in vitro* propagation of apple rootstocks to the propagation *in vitro* of the plum rootstock 'Pixy' (*Prunus insititia*) and the cherry rootstock Mazzard F12/1 (*P. avium*). Both of these rootstocks were in short supply, since both were difficult to propagate by conventional means (Howard, 1978; Feucht and Dausend, 1976). Results obtained from both cultivars indicated that some 6000 shoots could be produced from a single shoot apex over a period of 8 months. James and Thurbon (1979), employed similar *in vitro* techniques clonally to propagate the apple rootstock M9. This rootstock is widely used because of its effects on precocity and the control of tree and fruit size (Rogers and Beakbane, 1957; Tubbs, 1967). However, as with the *Prunus* rootstocks, cuttings are extremely difficult to root using conventional methods; softwood and hardwood cuttings give rooting percentages of 36 and 6 respectively (Howard, 1978). Subsequent reports by other workers extended such techniques to other apple rootstocks such as M7 (Werner and Boe, 1980) and also the MM series (Snir and Erez, 1980).

Following a report that apple plants could be regenerated from shoot meristem tips of certain apple scion cultivars (Lane, 1978), Jones *et al.* (1979) showed that the rapid clonal multiplication of self-rooted virus-free apple scion cultivars was feasible on a large scale. Cultures of 5 scion cultivars could be multiplied 5-fold every 3 weeks for many months; then the resulting shoots could be rooted and subsequently transferred to soil with high levels of success (80 and 90 per cent respectively). Other reports of the rapid *in vitro* propagation of difficult-to-root apple scion cultivars were produced by workers such as Zimmerman (1978); Zimmerman and Broome (1980); Sriskandrajah and Mullins (1981) and Srikandrajah, Mullins and Nair (1982). Such techniques have also been applied to crabapple cultivars such as *Malus sieboldii* var. zumi (Singha, 1982). Such cultivars are of value as ornamental plants, as potential pollinators for commercial apple cultivars (Crassweller, Ferree and Nichols, 1980) and as virus indicators (Gilmer *et al.*, 1971).

Almost immediately after the first reports concerned with the experimental micropropagation of fruit-tree rootstocks appeared, several commercial ventures were initiated to provide large scale production of virus-free plants of the dwarfing apple rootstocks M27, and of the plum rootstock Pixy (*Prunus insititia*) (Jones, 1979). Concurrent with these rapid advances in the application of techniques of *in vitro* propagation on a commercial scale, many reports appeared in the literature detailing the micropropagation of other temperate fruit tree species. The Round Table Conference of 1978, concerned with the *in vitro* multiplication of woody species, listed many species of fruit tree which were under investigation with a view to large scale clonal propagation. Since then, the techniques of rapid *in vitro* propagation have been successfully applied to many important members of the temperate zone tree fruits. Such reports have described the *in vitro* propagation of pear (Lane, 1979; Cheng, 1980; Singha, 1980), plum scion cultivars (Rosati, Marino and Swierczewski, 1980), almond scion cultivars (Rugini and Verma, 1982), scion cultivars of sweet cherry (Snir, 1982), peach (Miller, Coston, Denny and Romeo, 1982) and woody berry fruits such as blackberry and blueberry (Zimmerman, 1978). Other investigators in this area have also applied such techniques to provide clonal plantlets for research purposes (Jones, 1979) or to provide initial stocks of virus free plants. Huang and Millikan (1980), successfully applied the technique of *in vitro* micrografting of shoot tips to obtain apple plants free from apple stem grooving virus (SGV). Although most cultivars of apple show no symptoms when infected by this disease, the sensitivity of other cultivars has resulted in international quarantines against the sale of plants that are not certified as free from infection. Since this virus cannot be eliminated by thermotherapy, the use of *in vitro* techniques was the only practical alternative. Such *in vitro* grafting techniques are also proving to be of value in specific areas of research such as investigating scion/stock interactions (Negueroles and Jones, 1979).

Although *in vitro* propagation techniques are now well established for the rapid clonal multiplication of temperate fruit-tree species, it will be noted that all such reports focus on the culture of organised structures such as apical meristems or larger shoot tips. Multiple shoot proliferation is then achieved by inducing the outgrowth of axilliary meristems. There is a paucity of reports in the literature concerning either the induction of adventitious shoot production or organogenesis from other explant sources or from callus or cell cultures. Such limited reports have been confined mainly to studies with certain *Prunus* species, such as *P. amyadalus* (Mehra and Mehra, 1974), *P. mahaleb* (Hedtrich, 1977) and the hybrid *P. avium* x *pseudocerasus* (Gayner, Jones, Watkins and Hopgood, 1979). Other workers have reported somatic embryogenesis and adventitious bud production from tissues of grape (*Vitis* spp.) (Mullins and Srinivasan, 1976; Krul and Worley, 1977; Favre,

1977; and Barlass and Skene, 1978) and mulberry (*Morrus alba*) (Oka and Ohyama, 1981). Shih-Kin *et al.* (1977) reported the induction of callus from apple endosperm, and the subsequent differentiation of plantlets.

It is now generally accepted that rates of clonal multiplication made possible by the use of proliferation shoot-tip cultures are more than adequate for commercial application. Furthermore, even if it was possible to achieve plant regeneration from single cell cultures of fruit trees, the theoretically massive rates of multiplication made possible by such systems would be rarely attainable in practice. Also, such culture systems have the disadvantage of being inherently genetically unstable.

Although *in vitro* propagation techniques have been applied to most genera of the temperate zone tree fruits, the process of plant regeneration even from such an inherently simple organ culture system involves a complex sequence of growth stages (Wilkins and Dodds, 1982a). The following section will thus describe in detail the procedural steps necessary to regenerate clonal plantlets from meristem-tip and shoot-tip derived cultures of woody fruit plants.

Methodology of Micropropagation

The following discussion of the general methodology of fruit-tree micropropagation is based primarily on procedures presently in use in Birmingham, and developed from techniques cited by early investigators. In addition, this section will also highlight some of the cultural problems encountered both in this laboratory, and by other workers during the application of such techniques. Each of the procedural steps will be discussed in turn.

Surface Sterilisation. Vegetative shoots of the desired tree are surface sterilised. This process is usually accomplished by exposing the shoots to a solution of either calcium or sodium hypochlorite for a predetermined period of time. Although solutions of other surface sterilants have been tried (e.g. mercuric chloride, isopyropanol, hydrogen peroxide, etc.) hypochlorite solutions are normally recommended, since they are highly effective, but not as toxic as some of the other sterilants. Also, some workers have recommended that shoots be given a brief pre-rinse with ethanol prior to surface sterilisation. A drop of detergent (i.e. Tween 80) is normally added to the sterilising solution to assist as a wetting agent. Alternatively, dilute solutions of various commercial household bleaches may be used as sterilants. Due to the toxicity of most of the commonly used sterilants, shoots must be rinsed thoroughly at least 3 times with sterile distilled water prior to dissection. Once sterile, all subsequent culture operations are carried out under aseptic conditions, usually with the aid of a laminar flow cabinet. The use of a binocular dissecting microscope is usually necessary during the dissection process. Several authors have recommended specific treatments to minimise tissue damage as a result of the surface sterilisation process. Jones *et al.* (1977) recommend a two-step sterilisation process with an intermediate culture phase when initiating cultures of the M26 apple rootstock. Similarly, Zimmerman and Broome (1980) found that a short preculture period using liquid medium for 2-4 days, gave superior growth of cultures and lower explant losses. The use of liquid medium allows the leaching out of phenolic compounds produced by tissues damaged during the surface disinfestation procedure.

Culture Initiation. Cultures are usually initiated by dissecting away the shoot tip until only the apical dome, together with one or two leaf primordia, remain; the exposed meristem is then excised and placed onto the surface of a suitable culture medium. Most workers have found that a suitable culture medium for establishing shoot apices of most woody fruit species consists of the mineral salts and vitamins of either Murashige and Skoog (1962) or Linsmaier and Skoog (1965), supplemented with low concentration of a cytokinin (usually benzylamino purine (BAP) at 0.5 mg/l) together with 2-3 per cent (w/v) of sucrose as a carbon source. The use of other culture media has also been specified, for example Rugini and Verma (1982) have recommended the use of Tabachnik and Kester (1977) medium for initiating cultures of almond (*Prunus amygdalus* cv 'Ferragnes'). Also, for some species of pear (*Pyrus communis*, *P. pasha*) a higher concentration of BAP (2 mg/l) has been used (Wilkins and Dodds, 1982a). The initiation medium is usually solidified with purified agar at a concentration of 0.5-0.8 per cent (w/v).

The propensity of a particular meristem or shoot apex to form a culture can be dependent on many factors, these can include:

(1) The age and physiological status of the tree: Jones (1978), compared the ease of initiation of cultures from both 1-year-old and 8-year-old trees. He found that whereas shoot apices from 1-year-old trees were easy to establish, for 8-year-old trees there existed a 'critical explant size' of 3 mm, and it was impossible to initiate cultures from shoot apices of less than this size. Also, there was often a long delay before shoot-tips from the older trees would commence growth. Meristems are usually dissected in spring, just after bud-break, since in late

summer they become progressively more difficult to establish in culture. Jones (1978) has attempted to standardise the stage of development of material to be used as an explant source. Material is harvested whilst still dormant, then kept in a cold store at –3°C in plastic bags of damp peat with a fungicide, until required. Explants are then dissected just after bud-break, and little variation is observed if these techniques are applied.

(2) The size of the explant used to initiate the culture: some authors have reported that certain *Prunus* species (e.g. *P. domestica* cv Victoria) will not grow from meristems, but will grow from a shoot tip (Constantine, 1978). Similarly, Vertesy, Jones and Hopgood (1980) found that meristem tips could not be established from greenhouse grown plants of the plum rootstock Pixy (*Prunus insititia*), however once cultures were established from large shoot tips, meristems (0.1-0.2 mm) dissected from *in vitro* cultured shoots proved relatively easy to grow. Since, for purposes of virus eradication, only true meristems (0.1-0.2 mm) may be used, the use of such a two-step procedure may therefore facilitate virus eradication in those species whose meristems are difficult or impossible to grow.

(3) The position of the bud on the tree or branch: Lane (1978) reported that meristem tips of apple excised from near the terminus of a branch grew much better than meristem tips excised from near the basal buds.

Perhaps one of the most troublesome problems affecting the initiation of cultures of woody fruit plants is the phenomenon of explant browning and subsequent explant death. This is usually due to phenolic compounds produced by tissues damaged either by surface sterilisation, or during the dissection process. Several solutions to this problem have been proposed, such as the inclusion in the culture medium of compounds such as cysteine, ascorbic acid, dithiothreitol (DTT), polyvinylpyrrolidone (PVP) and activated charcoal. The use of a brief preculture period in agitated liquid medium has also been recommended, however no generally applicable solution to this problem has yet been found, although limiting the period of exposure to the surface sterilant, together with a swift, 'clean' dissection may help to alleviate the problem.

Culture Establishment. Initial cultures are normally incubated for 3-4 weeks (Figure 6.1 a, b) under high intensity illumination with a 16-hour photoperiod, although a continuous photoperiod has also been recommended. The culture temperature used is generally within the range of

Figure 6.1: Stages in the Establishment of Shoot Cultures of Various Temperate Fruit Trees: (a) shoot culture of the bird cherry (*Prunus padus*) three weeks after initiation, (b) shoot culture of common pear (*Pyrus communis*), three weeks after initiation, (c) established proliferating shoot culture of the 'EMLA' virus-free apple scion cultivar 'Greensleeves' (*Malus domestica*), (d) established proliferating shoot culture of the plum scion cultivar Jefferson's Gage (*Prunus domestica*)

22-28°C, dependent on the particular species being cultured. Shoots of various apple species and cultivars (*Malus* spp.) and some cherries (*Prunus* spp.) are cultured at 22-25°C, whilst shootcultures of pear species (*Pyrus communis*, *P. pashia*), plum scion and rootstock cultivars (*Prunus* spp.), black mulberry (*Morus nigra*) and pomegranate (*Punica granatum*) are grown at 28°C.

The volume of medium used for initiation of cultures is usually small (5-10 ml) as is the volume of the culture vessel used during the initiation phase (30-50 ml). This is necessary in order to maintain a high humidity within the culture vessel.

Proliferation. Once an actively growing culture is established, it may be transferred (without subdivision) to a larger culture vessel containing an increased volume (30-100 ml) of the same culture medium as before, except that the concentration of BAP is normally increased to either 1 or 2 mg/l, depending on the particular species or cultivar being cultured.

This technique has now been employed to establish cultures from a wide range of temperate fruit tree species at Birmingham, although problems have been encountered during the first transfer of the culture from the initiation phase to the establishment phase. This has been a problem with species such as *Malus sikkimensis*, *M. platycarpa*, *Pyrus longipes* and *Prunus cerasifera*. Rugini and Verma (1982) recommend the use of several different culture media during separate initiation, establishment and proliferation stages for the *in vitro* culture of the difficult-to-propagate almond cultivar 'Ferragnes' (*Prunus amygdalus*). It is possible that the application of such a multistage culture system may facilitate the establishment of cultures of the above listed species.

Once proliferating shoot cultures have been established (Figure 6.1 c, d), it is necessary to divide up individual shoot clusters at 3-5 week intervals and transfer portions, each with 2-3 shoots, to fresh proliferation medium. In this way, for most cultivars, a 4-20-fold shoot multiplication rate may be achieved, and may be maintained indefinitely.

The inclusion in the proliferation medium of phloroglucinol or other phenolic compounds such as pyrogallol, has been recommended by workers such as Jones (1976), Jones and Hopgood (1979), James (1978) and James and Thurbon (1979, 1981) as promoting shoot and root initiation in cultures of various apple and raspberry cultivars, and in the plum rootstock Pixy. However, other workers have found no benefits from the use of phloroglucinol, although the bacteriostatic action of this compound may be of some value.

With shoot cultures of most apple, plum and cherry cultivars, satisfactory rates of proliferation may be maintained simply by the division of shoot clumps, and subsequent transfer to fresh culture medium. However, the particular growth characteristics of some shoot cultures mean that a different approach must be used in order to maintain a satisfactory rate of shoot growth and proliferation. Lane (1979) found that proliferation in shoot cultures of pear (*Pyrus communis* L. cv 'Bartlett') could be optimised by shoot-tip removal followed by orientation of the shoot in either a horizontal or inverted position on the culture medium, in order to decrease apical dominance. In addition, a subculture period of 2-3 weeks is necessary for pear shoots, otherwise 3-4 isolated shoots become dominant, and subsequently inhibit the growth of smaller shoots. This phenomenon has been observed in several pear species (Wilkins and Dodds, unpublished observations). Similarly, established shoot cultures of pomegranate (*Punica granatum*), will only produce 2-4 shoots per 4-week culture period. However, since each shoot consists of 5-7 nodes, multiplication rates of 20-25-fold have been maintained by the simple transfer of single nodal cuttings (Wilkins and Dodds, unpublished observations).

One of the first requirements when establishing a system for the rapid clonal multiplication of fruit trees via *in vitro* techniques, is the need to optimise the conditions necessary to achieve maximum proliferation of vigorous shoots from cultures. Many workers have noted that rates of proliferation can vary even between cultivars of the same species. An experiment carried out in this laboratory, compared the rates of shoot proliferation in cultures of several rootstock and scion cultivars of apple. Cultures were grown in 150 x 25 mm test-tubes, each with 12 ml of a standard culture medium consisting of Murashige and Skoog (1962) salts and vitamins, supplemented with 3 per cent sucrose, 0.8 per cent purified agar and 1.0 mg/l BAP. Cultures were initiated from single shoots, and subsequently incubated for 4 weeks, under warm-white fluorescent tubes, with a 16-hour photoperiod. This mean proliferation rate (25 replicates) obtained for each cultivar ranged from 3 to 12 (Figure 6.2). Similar results have been observed with different *Prunus* cultivars (Wilkins and Dodds, unpublished results). Previous investigations (Wilkins and Dodds, 1982b) have shown that even establishing the optimum condition for growth and proliferation in shoot cultures of a single cherry cultivar can involve a considerable amount of painstaking empirical work.

Major differences can be observed in cultures grown using various culture systems, such as stationary liquid medium, agitated liquid medium or agar-solidified medium. Cultures grown using agitated liquid medium usually display better rates of growth due to increased adsorption of hormones and nutrients from the medium (Snir and Erez, 1980), or because agar may contain components which are toxic to some plant tissues (Kohlenbach and Wernicke, 1978). In addition, fruit tree cultures grown on agar must be transferred to fresh culture medium at regular intervals, otherwise browning of the medium occurs, followed by rapid senescence of the culture. This phenomenon is not observed with liquid culture systems.

Figure 6.2: Comparison of Rates of Shoot Proliferation Within Cultures of Various Apple Rootstock (M7, M9, M25, M27) and scion (Wijcik (Wij), Greensleeves (GS)) Cultivars. For details see text

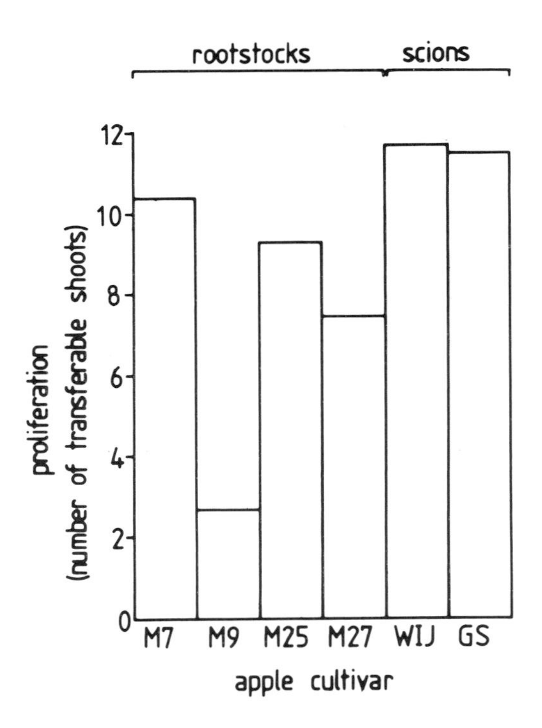

In the last 3-4 years, much interest has been shown in the physiological phenomenon of 'vitrification', also called hyperhydric transformation, glauciness, waterlogging and translucency (Debergh, 1981). The

phenomenon has been observed in many different fruit tree cultures, as well as cultures of many herbaceous plants. However, it seems to be especially common in shoot cultures of the apple rootstock M7, where the symptoms have been termed 'BA (benzyladenine) toxicity' (Werner and Boe, 1980). Several solutions to this problem have been proposed by various investigators, with varying degrees of success. Boxus, Druart and Brasseur (1978) were able to reduce the frequency of vitrification in their cultures of *Prunus* and *Malus* by placing cultures in a cold room at 3-4°C for a period of 3-4 weeks, together with the omission of BAP from the culture medium and leaf removal. Similarly, Rugini and Verma (1982) showed that vitrified shoots of almond (*Prunus amygdalus* Batsch. cv 'Ferragnes') could be restored to normality by keeping vitrified cultures at 4°C for 15 days. The same authors also observed that partial vitrification occurred in cultures grown on medium solidified with 0.7 per cent agar, but that a substantial improvement occurred on transferring the shoots to a medium containing 0.9 per cent agar. Furthermore, the inclusion in the culture medium of 0.5 per cent pectin together with 0.7 per cent agar allowed vigorous shoots to be generated repeatedly without problems of vitrification. Werner and Boe (1980) have recommended the periodic substitution of 5 mg/l N^6 (2 isopentyl) adenine (2iP) for BAP to prevent vitrification in cultures of M7 apple rootstock. A similar technique has been applied in this laboratory to cure problems of vitrification in shoot cultures of the plum rootstock Pixy (*Prunus insititia*).

Rooting. A great deal of attention has been given in the past to problems concerning the *in vitro* rooting and subsequent *in vivo* establishment of plantlets derived from fruit-tree shoot cultures. The system presently in use at Birmingham involves the exposure of shoots to a medium with a high auxin concentration (usually 3 mg/l indol-3yl-butyric acid (IBA)), for a specific period of time (usually 3-5 days). Shoots are then transferred to a dilute culture medium (10-50 per cent) with a reduced supply of carbohydrate (1-2 per cent sucrose), and devoid of growth regulators. Prolific extension of root initials usually occurs after 15-25 days incubation at 28°C (Figure 6.3a).

This two-step rooting procedure is usually necessary in order to allow root extension, since the inhibitory effect of IBA on root elongation is well known (West and Thimann, 1937; Thimann, 1977), and also the continued presence of auxin leads to excessive callusing at the base of the shoot (James and Thurbon, 1979). However, some cultivars of apple (e.g. the invigorating rootstock M25) root readily in the continued

Figure 6.3: Examples of the Types of Culture System Used for the in vitro *Rooting of Temperate Fruit Trees:* (a) selection of plantlets of the apple scion cultivar Wijcik (*Malus domestica*), rooted on agar-solidified medium; (b) selection of plantlets of the semi-dwarfing cherry rootstock Colt (*Prunus avium* x *pseudocerasus*) rooted using liquid medium with filter-paper strips as support.

a

b

presence of 0.2 mg/l IBA, and certain cherry rootstock cultivars, i.e. Colt (*Prunus avium* x *pseudocerasus*), will give 100 per cent rooting in the continued presence of 3 mg/l IBA, with little callusing (Figure 6.3b).

Precisely defined concentrations, both of the culture medium and the auxin used during the root initiation process are essential in order to avoid excessive callusing at the base of the plantlet, which can result in problems during *in vivo* establishment.

The auxins IBA and IAA (3-indoleacetic acid) are normally the preferred auxins for inducing rooting in shoots of apple, cherry and plum cultivars; whilst NAA (naphthalene acetic acid) is the preferred auxin for the rooting of pear shoots. The use of NAA in place of IBA or IAA for rooting pear shoots has been found to be preferable (Lane, 1979), since the use of NAA gives rise to short, thick roots, which are not easily damaged upon transfer to soil; root elongation then commences immediately the shoots are transferred to soil.

Ruigini and Verma (1982) recommend the use of a specific 'shoot elongation stage', to facilitate rooting of a difficult-to-propagate almond cultivar. Shoots are transferred from the proliferation medium onto a different culture medium containing a greatly reduced concentration of BAP (0.2 mg/l). Shoots then elongate over several days, forming 3-4 nodes; these shoots are then rooted by a two-step process of root initiation and elongation. Root initiation is achieved by exposure to 1 mg/l of either IBA or IAA, whilst root elongation is achieved by transferring to liquid medium with sterile vermiculite as a support, and devoid of growth regulators. The use of liquid medium has been recommended as being beneficial for rooting, since it helps to avoid contamination problems sometimes encountered when agar-rooted plantlets are transferred to non-sterile conditions. Since rooting plantlets usually have to be supported in liquid medium, various substrates have been proposed for this purpose, these include: filter-paper bridges, vermiculite, perlite, rockwool, cellulose fibres, and synthetic cellulose polymers. The most efficient substrates are usually those that effectively minimise root damage when transferring the plantlet to non-sterile conditions. Werner and Boe (1980), specified the use of semi-liquid agar at a concentration of 0.27 per cent for root elongation, to minimise root damage on transfer.

In order to optimise root proliferation, some workers recommend wounding the base of the shoot prior to root initiation; this has been tried successfully on apple rootstock cultivars (Snir and Erez, 1980) and sweet cherry cultivars (Snir, 1982). It has also been reported that with some species the duration of culture can greatly affect the rooting percentages. Sriskandarajah *et al.* (1982) found that with the apple scion cultivar Delicious, there was a progressive improvement in the rooting of shoots with increasing numbers of subcultures. After 4 subcultures, the percentage rooting was 21, whilst after 31 subcultures, the percentage rooting had risen to 79.

Plantlet Establishment. Once a satisfactory root system has formed, plantlets are transferred to trays of sterile loamless compost in a glasshouse or growth chamber, and grown under high intensity illumination at 20-25°C (Figure 6.4a). The use of a plastic propagator with a simple venting system, or plastic sheeting, is necessary to maintain conditions of high humidity and avoid dessication problems. After 7-10 days, plantlets will have hardened off sufficiently to be transferred to individual pots (Figure 6.4b), and then grown-on prior to transferring to field conditions. During the establishment period, plantlets must be fed regularly with a nutrient solution; usually a 10 per cent solution of the major mineral salts of Murashige and Skoog (1962). Several variations of this system have been proposed by various other investigators. Most of these alternative schemes have employed intermittent mist application to maintain conditions of high humidity, together with the substitution of various rooting substrates for loamless compost. Several workers have recommended the use of various mixtures of sand, sphagnum peat moss, perlite, vermiculite and potting soil in which to establish plantlets. However, it would appear that the only critical factor in such a choice is the use of a well-drained rooting medium. The application of a fungicide treatment has also been specified by various investigators. However, problems still remain concerning the establishment of *in vitro* rooted microcuttings and it is highly likely that future research will be aimed at the development of a generally applicable system which can be routinely applied to a wide range of temperate fruit tree species.

Conclusions

New reports are continually appearing in the literature concerning the successful propagation of fruit tree species by the application of *in vitro* techniques. Also, in spite of the high capital cost of establishing a tissue-culture facility, such techniques are finding increasing commercial application for the rapid clonal multiplication of desirable fruit-tree cultivars, and it is likely that this trend will continue in the future.

Figure 6.4: Stages in the Establishment of Rooted Plantlets of Fruit Trees Under Non-sterile Conditions: (a) tray containing plantlets of the cherry rootstock Colt (*Prunus avium* x *pseudocerasus*). two weeks after transfer to non-sterile conditions, and (b) selection of clonal plantlets of Colt, after six weeks growth under greenhouse conditions.

a

b

References

Abbott, A.J. and Whiteley, E. (1976) 'Culture of *Malus* tissues *in vitro*. I. Multiplication of apple plants from isolated shoot apices', *Sci. Hort.*, *4*, 183-9

Barlass, M. and Skene, K.G.M. (1978) '*In vitro* propagation of grapevine (*Vitis unifera* L.) from fragmented shoot apices', *Vitis.*, *17*, 335-40

Boxus, P., Druart, P. and Brasseur, E. (1978) Activiteitsverslag Rijkscentrum voor Landbouwkundig Onderzoek, Gembloux, 124

Boxus, P. and Quoirin, M. (1977) 'Comportement en pepiniere d'arbes fruitiers issus de culture *in vitro*', *Acta. Hort.*, *78*, 373-8

Cheng, T.-Y. (1980) 'Micropropagation of Fruit Tree Rootstocks' in *Proceedings of the Conference on Nursery Production of Fruit Plants through Tissue Culture – Applications and Feasibility*, USDA, Science and Education Administration Agricultural Research Results, ARR-NE-11, Beltsville, Maryland

Constantine, D. (1978) '*In vitro* Multiplication of Woody Species' in *Round Table Conference*, C.R.A. Gembloux (Belgium), p. 5

Crassweller, R.M., Ferree, D.C. and Nichols, L.P. (1980) 'Flowering crab apples as potential pollinizers for commercial apple cultivars', *J. Am. Soc. Hort. Sci.*, *105*, 475-7

Debergh, P. (1981) 'Mass propagation of globe artichoke (*Cynara scolymus*): Evaluation of different hypotheses to overcome vitrification with special reference to water potential', *Physiol. Plant.*, *53*, 181-7

Dutcher, R.D. and Powell, L.E. (1972) 'Culture of apple shoots from buds *in vitro*', *J. Am. Soc. Hort. Sci.*, *97*, 511-14

Elliot, R.F. (1972) 'Axenic culture of shoot apices of apple', *N.Z. J. Bot.*, *10*, 254-8

Favre, J.-M. (1977) 'Premiers resultats concernant l'obtention *in vitro* de neoformations caulinaires chez la vigne', *Ann. Amelior. Plantes*, *27*, 151-69

Feucht, W. and Dausend, B. (1976) 'Root induction *in vitro* of easy-to-root *Prunus pseudocerasus* and difficult-to-root *P. avium*', *Scientia Hort.*, *4*, 49-54

Gayner, J.A., Jones, O.P., Watkins, R. and Hopgood, M.E. (1979) Report of East Malling Research Station for 1979, p. 187

Geissbuhler, H. and Skoog, F. (1957) 'Comments on the application of plant tissue cultivation to propagation of forest trees', *Tappi*, *40*, 258-62

Gilmer, R.M., Mink, G.I., Shay, J.R., Stouffer, R.F. and McCrum, R.C. (1971) 'Latent viruses of apple. I. Detection with woody indicators', *Search (Agr)*, I(*10*), 1-21. N.Y. State. Agr. Expt. Station

Haissig, B.E. (1965) 'Organ formation *in vitro* as applicable to forest tree propagation', *Bot. Rev.*, *31*, 607-26

Hedtrich, C.M. (1977) 'Differentiation of cultivated leaf discs of *Prumus mahaleb*', *Acta. Hort.*, *78*, 177-83

Howard, B.H. (1978) 'Propagation and nursery production', Report of the East Malling Research Station for 1977, pp. 67-71

Huang, S.-C. and Millikan, D.F. (1980) 'In vitro micrografting of apple shoot tips', *Hort. Sci.*, *15*(6), 741-3

Hudson, J.P. (1971) 'Meadow Orchards', *Agriculture*, *78*, 157-60

Hudson, J.P. (1982) 'New perspectives in vegetative propagation', *Outlook on Agriculture*, *11*(2), 55-61

James, D.J. (1978) Report of East Malling Research Station for 1977, p. 177

James, D.J. and Thurbon, I.J. (1979) 'Rapid *in vitro* rooting of the apple rootstock M.9', *J. Hort. Sci.*, *54*(4), 309-11

James, D.J. and Thurbon, I.J. (1981) 'Shoot and root initiation in vitro in the apple rootstock M.9. and the promotive effects of phloroglucinol', *J. Hort. Sci.*, *56*(1), 15-20

Jones, O.P. (1967) 'Effect of benzyladenine on isolated apple shoots', *Nature*, *215*, 1514-15

Jones, O.P. (1976) 'Effect of phloridzin and phloroglucinol on apple shoots', *Nature*, *262*, 392-3. Erratum, id., *262*, 724

Jones, O.P. (1978) '*In vitro* Multiplication of Woody Species' in *Round Table Conference*, C.R.A. Gembloux (Belgium), p. 22

Jones, O.P. (1979) 'Propagation *in vitro* of apple trees and other woody fruit plants', *Sci. Hort.*, *30*, 44-8

Jones, O.P. and Vine, S.J. (1968) 'The culture of gooseberry shoot tips for eliminating virus', *J. Hort. Sci.*, *43*, 289-92

Jones, O.P., Hopgood, M.E. and O'Farrell, D. (1977) 'Propagation in vitro of M26 apple rootstocks', *J. Hort. Sci.*, *52*, 235-8

Jones, O.P. and Hopgood, M.E. (1979) 'The successful propagation in vitro of two rootstocks of Prunus: the plum rootstock Pixy (P. insititia) and the cherry rootstock F12/1 (*P. avium*)', *J. Hort. Sci.*, *54*, 63-6

Jones, O.P., Pontikis, C.A. and Hopgood, M.E. (1979) 'Propagation in vitro of five apple scion cultivars', *J. Hort. Sci.*, *54*, 155-8

Kohlenbach, H.W. and Wernicke, W. (1978) 'Investigations on the inhibitory effect of agar and the function of active carbon in another culture', *Z. Pflanzenphysiol.*, *86*, 463-72

Krul, W.R. and Worley, J.F. (1977) 'Formation of adventitious embryos in callus cultures of "Seynal" a French hybrid grape', *J. Am. Soc. Hort. Sci.*, *102*, 360-3

Lane, W.D. (1978) 'Regeneration of apple plants from shoot meristem-tips', *Plant Sci. Lett.*, *13*, 281-5

Lane, W.D. (1979) 'Regeneration of pear plants from shoot meristem-tips', *Plant Sci. Lett.*, *16*, 337-42

Linsmaier, E.M. and Skoog, F. (1965) 'Organic growth factor requirements of tobacco tissue cultures', *Physiol. Plant.*, *18*, 100-27

Mehra, A. and Mehra, P.N. (1974) 'Organogenesis and plantlet formation *in vitro* in Almond', *Bot. Gaz.*, *135*, 61-73

Morel, G.M. and Martin, C. (1955) 'Guerison de plantes atteintes de maladies à virus par culture de meristemes apicaux', *Rep. XIVth. Int. Hort. Cong.*, *1*, 303-10

Miller, G.A., Coston, D.C., Denny, E.G. and Romeo, M.E. (1982) '*In vitro* propagation of 'Nemaguard' peach rootstock', *Hort. Sci.*, *17*(2), 194

Mullins, M.G. and Srinivasan, C. (1976) 'Somatic embryos and plantlets from an ancient clone of the grapevine (cv. Cabernet-Sauvignon) by apomixis *in vitro*', *J. Exp. Bot.*, *27*, 1022-30

Murashige, T. and Skoog, F. (1962) 'A revised medium for rapid growth and bioassay with tobacco tissue cultures', *Physiol. Plant.*, *15*, 473-97

Negueroles, J. and Jones, O.P. (1979) 'Production *in vitro* of rootstock/scion combinations of *Prunus* cultivars', *J. Hort. Sci.*, *54*, 279-81

Oka, S. and Ohyama, K. (1981) '*In vitro* initiation of adventitious buds and its modification by high concentration of benzyladenine in leaf tissues of mulberry (*Monus alba*)', *Can. J. Bot.*, *59*(1), 68-74

Pieriazek, J. (1968) 'The growth in vitro of isolated apple-shoot tips from young seedlings on media containing growth regulators', *Bull. Acad. Pol. Sci., Ser. Sci. Biol.*, *16*, 179-83

Pierik, R.L.M. (1975) 'Vegetative propagation of horticultural crops in vitro with special attention to shrubs and trees', *Acta. Hort.*, *54*, 71-82

Quoirin, M. (1974) 'Premiers resultats obtenus dans la culture in vitro du meristeme apical du sujets porte-greffe du pommier', *Bull. Rech. Agronom. Gembloux*, *9*, 189-92

Rogers, W.S. and Beakbane, A.B. (1957) 'Stock and scion relations', *Ann. Rev. Plant. Physiol.*, *8*, 217-36

Rosati, P., Marino, G. and Swierczewski, C. (1980) '*In vitro* propagation of Japanese plum (*Prunus salicina* Lindl. cv. Calita)', *J. Am. Soc. Hort. Sci.*, *105*, 126-9

Round Table Conference (1978) In vitro *multiplication of woody species*, C.R.A. Gembloux, Belgium

Rugini, E. and Verma, D.C. (1982) 'Micropropagation of difficult to propagate almond (*Prunus amygdalus*) cultivar', *Plant. Sci. Lett.* (in press)

Sekowski, B. (1956) *Pomologia*, Poznan

Shih-Kin, M., Shu-Qiong, L., Yue-Kun, Z., Nan-Fen, Q., Peng, Z., Hong-Xun, X., Fu-Shou, Z. and Zhen-Long, Y. (1976) 'Induction of callus from apple endosperm and differentiation of the endosperm plantlet', *Scientia Sinica.*, *XX*(3), 370-5

Singha, S. (1980) '*In vitro* propagation of 'Sekel'', *Pear. Proc. Conf. Nursery Production of Fruit Plants through Tissue Culture – Applications and Feasibility*, U.S. Dept. Agr. SEA. ARR-NE-11, pp. 59-63

Snir, I. (1982) '*In vitro* propagation of sweet cherry cultivars', *Hort. Sci.*, *17*(2), 192-3

Snir, I. and Erez, I. (1980) '*In vitro* propagation of Malling Merton apple rootstocks', *Hort. Sci.*, *15*, 597-8

Sriskandarajah, S. and Mullins, M.G. (1981) 'Micro-propagation of Granny Smith apple: factors affecting root formation *in vitro*', *J. Hort. Sci.*, *56*(1), 71-6

Sriskandarajah, S., Mullins, M.G. and Nair, Y. (1982) 'Induction of adventitious rooting *in vitro* in difficult-to-propagate cultivars of apple', *Plant Sci. Lett.*, *18*, pp. 1-9

Tabachnik, L. and Kester, D.E. (1977) 'Shoot culture for almond and almond-peach hybrid clones *in vitro*', *Hort. Sci.*, *12*, 545-7

Thimann, K.V. (1977) 'The initiation of roots on stems' in *Hormone action in the whole life of the plant*, University of Massachusetts Press, Amherst, pp. 190-205

Tubbs, F.R. (1967) 'Tree size control through dwarfing rootstocks', *Proc. XVII Int. Hort. Cong.*, *3*, 43-56

Tukey, H.B. (1964) *Dwarfed fruit trees*, Macmillan, New York

Vavilov, N.I. (1926) *Studies on the Origin of Cultivated Plants*, Leningrad

Vertesy, J., Jones, O.P. and Hopgood, M.E. (1980) Report of the East Malling Research Station for 1979, p. 187

Walkey, D.G. (1972) 'Production of apple plantlets from axillary-bud meristems', *Can. J. Plant. Sci.*, *52*, 1085-7

Werner, E. and Boe, A.A. (1980) '*In vitro* propagation of Malling 7 apple rootstock', *Hort. Sci.*, *15*, 509-10

West, F.W. and Thimann, K.V. (1937) 'Root formation' in *Phytohormones*, Macmillan, New York

Westwood, M.N. (1978) *Temperate Zone Pomology*, W.H. Freeman and Company, San Francisco

Wilkins, C.P. and Dodds, J.H. (1982a) 'The application of tissue culture techniques to plant genetic conservation', *Sci. Prog.* (Oxf.), *68*, 281-307

Wilkins, C.P. and Dodds, J.H. (1982b) 'The *in vitro* growth of cherry root tips in the presence of various plant growth regulators', *Plant Growth Regulation*, *1*, 209-16

Zagaja, S.W. (1970) 'Temperate zone tree fruits' in O.H. Frankel and E. Bennet (eds.), *Genetic Resources in Plants – their Exploration and Conservation*, IBP Handbook, no. 11, Blackwell Scientific Publications, Oxford, pp. 327-33

Zielinski, Q.B. (1955) *Modern Systematic Pomology*, Iowa State University Press, Ames, Iowa

Zimmerman, R.H. (1978) 'Tissue culture of fruit trees and other fruit plants', *Proc. Intern. Plant. Prop. Soc.*, *28*, 539-45

Zimmerman, R.H. and Broome, O.C. (1980) 'Apple Cultivar Micropropagation' in *Proc. Conf. Nursery Production of Fruit Plants through Tissue Culture – Applications and Feasibility*, U.S. Dept. Agr. – SEA. ARR-NE-11, pp. 54-8

7 TISSUE CULTURE OF CITRUS

John H. Dodds

Citrus are all thought to be native to the tropical and subtropical regions of Asia and the Malay Archipeligo (Webber, 1967). Citrus is now however a major crop and Table 7.1 shows the output of the major citrus producing countries for 1981. Oranges and tangerines account for well over three-quarters of total citrus production, the balance being mainly grapefruits, lemons and limes.

Table 7.1: Table of Citrus*-producing Countries*

Country	Production (tonnes × 1000) 1981
USA	13 500
Brazil	5 600
Japan	5 000
Spain	3 200
Italy	2 800
Mexico	2 310
India	2 000
Argentina	1 950
Egypt	1 730
Israel	1 600
Morocco	1 570
China	1 220
Others	11 400
World Total	53 870

In commercial propagation of citrus the normal fruit tree problem of lack of uniformity of seedlings does not arise as in *Citrus* only highly polyembryonic clones are selected for rootstock seedlings. Many *Citrus* species are polyembryonic and produce anything up to 40 adventive embryos to the nucellus (Furusatro, Onta and Ishibashi, 1957). The number of polyembryoids formed is dependent on the nutritional status of the fruit. In theory nucellar-produced trees will be identical to the mother tree (Frost and Soost, 1968).

It is believed to be unlikely that somatic fertilisation takes place in *Citrus*, however it is impossible to rule out the possibility that pollination and fertilisation could influence the gene expression regime in the nucellar embryos. Inglesias, Lima and Simon (1974) have also demonstrated some genetic inconsistency or variation between nucellar embryos.

Indirect Adventitious Embryo Induction

The initiation of embryos in callus formed from proliferation of unpollinated ovaries has been described by Mitra and Chaturvedi (1972). Kochba, Spiegal-Roy and Safran (1972) obtained an embryogenic callus from unfertilised ovules of the 'Shamouti' orange (*Citrus sinensis*). Stimulation of embryoid production was obtained if the callus was subjected to 16 krad of gamma-ray irradiation. This result could be obtained simply by irradiation of the medium and hence there must be generation of new embryogenic substances in the media (Spiegel-Roy and Kochba, 1973). Further studies of the callus embryo system (Bitters *et al.*, 1974) showed that the tissue produced globular embryos from single cells on the surface of the callus or on the surface of existing pro-embryo structures.

The ability to derive plantlets from single cells in the *Citrus* system has shown great promise for the development of protoplast and mutant studies, more of which will be covered later in this chapter.

Initiation and Culture of Vegetative Buds

Adventitious buds were induced in callus cultures derived from stem tissue of seedlings grown *in vitro* (Grinblat, 1972; Chaturvedi and Mitra, 1974). In both instances buds developed into normal shoots which were then transferred to a second medium for the induction of roots. As expected, benzyl adenine promoted bud formation but depressed or prevented rooting. Root formation was induced on a medium containing auxin. Marked qualitative and quantitative changes in amino acid composition of the cultures was correlated with bud formation (Chaturvedi, Chowdhury and Mitra, 1974b). Unfortunately the lack of sufficient anatomical and cytological detail prevented them from determining whether these changes were the cause or result of bud initiation.

Excised vegetative buds have been grown aseptically to study the effects of applied growth regulators (Altman and Goren, 1971, 1974a, b). It was found (1974b) that sprouting of buds *in vitro* was affected by regulators in a similar way to that of buds *in vivo*. Sprouting was retarded by IAA and ABA in the medium, but slightly enhanced by GA_3. The sprouting of numerous shoots from single explants was induced by BA and kinetin, while GA_3 caused the new growth to have excessively long internodes. The similarity in response between buds grown *in vivo* and *in vitro* is significant. It permits the study *in vitro* of factors controlling complex phenomena such as flower bud induction and development *in vivo*.

Problems with Seedless Citrus

Lack of seeds in *Citrus* is a commercially desirable trait but one which normally prevents the production of nucellar seedlings.

Bitters *et al.* (1970, 1972) derived virus-free seedlings of Robertson navel orange from nucellar isolates *in vitro*. Cultures were initiated from ovules immediately after pollination and for six weeks, when the best results were obtained. Beyond this period, complete ovule degeneration prevented the establishment of further cultures.

Button and Bornman (1971a) succeeded in inducing embryogenesis in nucellar explants and in nucelli within unfertilised ovules of Washington navel orange. Entire ovules were not only most easily obtained and handled, but yielded more embryos per culture than did nucellar isolates. Furthermore, embryogenesis in three navel orange clones occurred readily in ovules removed from fruits of up to 40 mm diameter (3-4 months post anthesis). Nucellar explants from such large fruits invariably failed to proliferate. Thus the use of entire, unfertilised ovules extends the period during which nucellar cultures can be successfully initiated.

More recently this technique has been used to import two navel orange clones into South Africa with very little danger of inadvertent virus importation. These clones have been vegetatively propagated on a number of rootstocks and are now in trial orchards. The development of plantlets from unfertilised ovules of potentially seeded, polyembryonic cultivars has also been reported (Kochba *et al.*, 1972; Mitra and Chaturvedi, 1972).

Meristem Tip Culture

Attempts to produce micropropagation systems from *Citrus* like those with other fruit trees (see Chapter 6) have so far not met with success.

It is possible however to keep entire buds in culture for many months (Altman and Goren, 1971). Techniques have been developed by Murashige (1972) and Navarro, Roistacher and Murashige (1975) to graft shoot tips (meristematic dome plus between three and six leaf primordia) onto seedling rootstocks. The technique has been used in a number of important *Citrus* clones to free the material from tristeza, exocortis and stubborn disease. This technique offers the advantage over nucellar cultures that the disease can be eradicated without the reappearance of juvenile tissues.

Protoplast Culture, Mutant Selection and Somatic Hybridisation

Since it had been shown previously that embryoids could be formed from callus cultures (Kochba and Spiegel-Roy, 1977; Kochba *et al.*, 1972) it was obviously highly desirable to devise technology for protoplast isolation and culture, with the aim of being able to regenerate whole plants from single isolated cells. Vardi, Spiegel-Roy and Galun (1975, 1976) and Vardi (1977) isolated protoplasts from *Citrus sinensis* (Shamouti orange) and cultured these until a callus was formed. It was then possible to induce formation of embryoids on the callus as shown in Figures 7.1 and 7.2. Figure 7.3 shows a range of young plantlets derived from protoplasts for a number of *Citrus* varieties (Vardi *et al.*, 1982).

Figure 7.1: Callus and Embryoids from a Cultured Ovule of 'Ponkan'. Torpedo embryoids (TE) are moved to reveal the ovular integument (OI) with attached callus and globular embryos (GE)

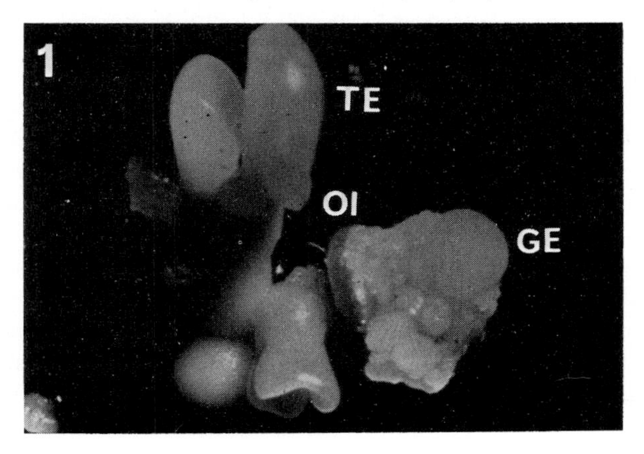

Figure 7.2: Nucellar Calli Derived from Cultured Ovaries Dissected from Young Fruits. (a) Dancy mandarin, (b) Ponkan mandarin

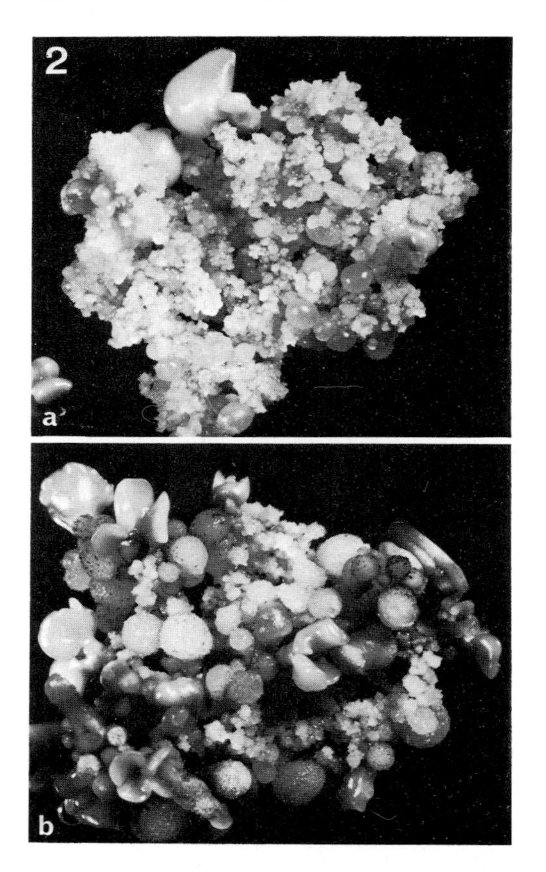

Vardi (1977) looked at the effect of mutagenic treatments such as 0.3 per cent ethyl methane sulphonate for one hour and X-ray radiation of about 3400R. Using these methods it was possible to regenerate whole mutant plants which can be included in screening studies.

It is interesting to note that polyploidisation is frequently observed in protoplast-derived plantlets (see Zelcer *et al.*, 1978). Some of the protoplast-derived plantlets obtained by Vardi (1978) were grafted onto sour orange and grew to fruiting stage; these were proven to be diploids. Another interesting feature is that *Citrus* protoplasts, unlike other plants, do not require an auxin in the culture media to initiate

Figure 7.3: Young Citrus Plants Derived from Cultured Protoplasts Photographed Five Months After Planting. (a) Orange (Shamouti), Mandarin (Murcott, Dancy, Ponkan) and Lemon (Villafranca) plants, (b) Plants derived from Dancy protoplasts – note uniformity

callus. This absence of auxin may be linked to the lack of polyploidisation in the tissue.

Potential for the Future

Techniques are already available for the regeneration of whole *Citrus* plants from isolated protoplasts, which means that the technology is already available for the selection and regeneration of mutants. This will include selection of resistance mutants to certain stress factors, such as high NaCl concentration.

It should also be possible in the future to study the formation of non-sexual hybrids of *Citrus* by somatic fusion techniques. The potential for hybrid formation by this method must hold great promise for the *Citrus* industry, although any successful product must be many years distant.

References

Altman, A. and Goren, R. (1971) 'Promotion of callus formation by ABA in *Citrus* bud cultures', *Plant Physiol.*, *47*, 844-6

Altman, A. and Goren, R. (1974a) 'Growth and dormancy cycles in *Citrus* bud cultures and their hormonal control', *Physiol. Plant*, *30*, 240-5

Altman, A. and Goren R. (1974b) 'Inter-relationship of ABA and GA in the promotion of callus formation in the abscission zone of *Citrus* bud cultures', *Physiol. Plant*, *32*, 55-61

Bitters, W.P., Murashige, T., Rangan, T.S. and Nauer, E. (1970) 'Investigations on established virus-free plants through tissue culture', *Calif. Citrus Nurgmans Soc.*, *9*, 27-30

Bitters, W.P., Murashige, T., Rangan, T.S. and Nauer, E. (1972) 'Investigations on establishing virus-free *Citrus* plants through tissue culture' in W.C. Price (ed.), *Proceedings of the 5th International Congress of the International Organ of Citrus Virology*, Gainesville, Florida, pp. 267-77

Button, J. and Bornman, C.H. (1971) 'Development of nucellar plants from unpollinated and unfertilised ovules of Navel orange *in vitro*', *J. South African Bot.*, *37*, 127-34

Button, J., Kochba, J. and Bornman, C.H. (1974) 'Fine structure of and embryoid development from embryogenic ovular callus of Shamouti orange', *J. Exp. Bot.*, *25*, 446-57

Chaturvedi, H.C. and Mitra, G.C. (1974) 'Clonal propagation of *Citrus* from somatic callus cultures', *Hort. Sci.*, *9*, 118-20

Chaturvedi, H.C., Chowdhury, A.R. and Mitra, G.C. (1974a) 'Shoot bud differentiation in stem callus tissue of *Citrus grandis* and correlated changes in its free amino acid content', *Current Sci.*, *43*, 536-7

Chaturvedi, H.C., Chowdhury, A.R. and Mitra, G.C. (1974b) 'Morphogenesis in stem callus tissue of *Citrus grandis* in long term cultures. A biochemical analysis', *Current Sci.*, *43*, 139-42

Frost, H.B. and Soost, R.K. (1968) 'Seed reproduction: development of gametes and embryos' in J. Reuther *et al.* (eds.), *The Citrus Industry*, Riverside, Univ. California, pp. 290-324

Furusatro, K., Onta, Y. and Ishibashi, K. (1957) 'Studies on polyembryony in *Citrus*', *Rep. Kihara Inst. Biol. Res.*, *8*, 40-8

Grinblat, U. (1972) 'Differentiation of citrus stem *in vitro*', *J. Am. Soc. Hort. Res.*, *97*, 599-603

Inglesias, L., Lima, H. and Simon, J.P. (1974) 'Isozyme identification of zygotic and nucellar seedlings in *Citrus*', *J. Hered.*, *65*, 81-4

Kochba, J. and Spiegel-Roy, P. (1973) 'Effect of culture media on embryoid formation from ovular callus of Shamouti orange', *Z. Pflanzenzucht*, *69*, 156-62

Kochba, J., Spiegel-Roy, P. and Safran, H. (1972) 'Adventive plant ovules and nucelli in *Citrus*', *Planta*, *106*, 237-45

Mitra, G.C. and Chaturvedi, H.C. (1972) 'Embryoids and complete plants from unpollinated ovaries and from ovules of *in vivo* grown emasculated flower buds of *Citrus* spp. *Bell*', *Torrey Bot. Cl.*, *99*, 184-9

Murashige, T. (1974) 'Plant propagation through tissue cultures', *Ann. Rev. Plant Physiol.*, *25*, 135-65

Navarro, L., Roistacher, C.N. and Murashige, T. (1975) 'Improvement of shoot tip grafting *in vitro* for virus free *Citrus*', *J. Am. Soc. Hort. Sci.*, *100*, 471-9

Spiegel-Roy, P. and Kochba, J. (1973) 'Stimulation of differentiation in orange (*Citrus sinensis*) ovular callus in relation to irradiation of the media', *Radiation Bot.*, *13*, 97-103

Vardi, A. (1977) 'Isolation of protoplast in *Citrus*', *Proc. Int. Soc. Citriculture*, *2*, 575-8

Vardi, A. (1978) 'Studies on isolation and regeneration of orange protoplasts' in A.W. Alferman and E. Reinhard (eds.), *Proc. Int. Symp. Plant Cell Culture*, pp. 234-42

Vardi, A., Spiegel-Roy, P. and Galun, E. (1975) '*Citrus* cell culture isolation of protoplasts, plating densities, effect of mutagens and regeneration of embryoids', *Plant Sci. Lett.*, *4*, 231-6

Vardi, A. and Raveh, D. (1976) 'Cross feeder experiments between tobacco and orange protoplasts', *Z. Pflanzenphysiol.*, *78*, 350-9

Vardi, A., Spiegel-Roy, P. and Galun, E. (1982) 'Plant Regeneration from *Citrus* Protoplasts. Variation in methodological requirements for various species', *Theor. Appl. Genet.*, *62*, 171-6

Webber, H.J. (1967) 'History and development of the *Citrus* industry' in J. Reuther *et al.* (eds.), *The Citrus Industry*, Riverside, Univ. California, pp. 1-39

Zelcher, A., Aviv, D. and Galun, E. (1978) 'Interspecific transfer of cytoplasmic male sterility by fusion between protoplasts of normal *Nicotiana sylvestris* and X irradiated protoplasts of male sterile *N. tabacum*', *Z. Pflanzenphysiol.*, *90*, 397-407

8 THE INFLUENCE OF PHYSICAL FACTORS ON XYLEM DIFFERENTIATION *IN VITRO*

Lorin W. Roberts

Temperature, water, light, gas composition, mechanical stress and acidity are the most important physical factors in the modification of tracheary element differentiation in cultured plant tissues. The significance of some of these factors in the initiation of cytodifferentiation was reviewed by Roberts (1976). Investigations on the possible effects of physical factors on tracheary element formation *in vitro* can be subdivided into three categories: cellular, callus and aggregate cultures in suspension, and regenerated plantlets. There has been little or no research conducted on single cells; most studies have examined the role of physical variables in callus cultures. Research in the third and most important category, i.e. the influence of physical factors on tracheary element production in regenerated plantlets, is still in its infancy. The leader in the latter field has been Murashige (1974), who devised the optimum cultural conditions for many species of plantlets during stage III, which involves the transition between *in vitro* and *in vivo* growth. Unfortunately, these studies on environmental factors have not been concerned with the optimal physical conditions for secondary differentiation of vascular tissues during the early growth period of these young plantlets.

The findings of Grout and Aston (1977) demonstrate the importance of further studies on vascular differentiation during organogenesis and plantlet formation. These workers found incomplete vascular connections between roots and shoots of regenerated cauliflower (*Brassica oleracea* var. botrytis cv. 'Armado Tardo') plantlets, which seriously restricted acropetal water movement and developmental growth. Because of the paucity of information on the relationship between physical factors on xylogenesis in plantlets, the present review will be restricted to the investigations on physical factors influencing xylem differentiation in primary explants and callus cultures. Denne and Dodd (1981) have reviewed the environmental control of xylem differentiation in trees.

In general, what are the effects of physical factors on the formation of tracheary elements? The critical variables or chemical necessities for the initiation of this cytodifferentiation event in a parenchymatous explant of quiescent cells include supplementing the basal medium with an auxin, a cytokinin, and a carbon source (Roberts, 1976; Savidge and Wareing, 1981); in addition there is growing evidence that ethylene is a hormone in this process (Roberts *et al.*, 1982). Since ethylene biosynthesis is readily initiated by any physical stress on the living system (Lieberman, 1979), the possibility of this occurring in a given experiment must be considered throughout the present review. Physical factors may influence cytodifferentiation by affecting either hormone activity or availability, the carbon source, or some metabolic pathway known to be associated with the developmental process. An example of the latter control is the effect of light on the regulation of phenylalanine ammonia-lyase activity (PAL). PAL is an important enzyme during the lignification stage of tracheary element differentiation.

Temperature

Most studies on the *in vitro* growth and development of plant cells indicate that the optimum temperature lies within the 25-30°C range, although some species differ considerably (Martin, 1980). The physiological effects of temperature extremes on plants has been reviewed recently by McDaniel (1982).

Low Temperature Effects

Durzan, Chafe and Lopushanski (1973) found that xylem differentiation occurred in suspension cultures of white spruce (*Picea glauca* [Moench] Voss) with constant light at 23.5°C, but xylem formation was completely blocked by growing the cultures under temperature conditions of a late-spring day. The latter inhibitory treatment consisted of cycles of 14 hr light (23.5°C) and 10 hr dark (12°C). Tannin biosynthesis by the cultures grown under the alternating environmental conditions may have blocked xylogenesis (Chafe and Durzan, 1973; Jacquoit, 1947).

Gautheret (1961) found that low temperatures (<17°C) were inhibitory to cytodifferentiation in callus of *Helianthus tuberosus*. Later Gautheret (1969) reviewed the effects of temperature and light on rhizogenesis and vascular differentiation in tuber explants of *Helianthus tuberosus*. A temperature of 26°C was necessary for the induction of vascular differentiation after four days of culture, and the formation of a cambium was evident following six days of growth at this temperature. Cultured tissues grown for 21 days at 15°C differentiated phloem elements but not tracheary elements; apparently the induction of cytodifferentiation of the two types of vascular tissue has different temperature requirements in this system (Gautheret, 1969).

Shininger (1979a, b) found that xylogenesis induced in excised and cultured pea (*Pisum sativum* cv. Little Marvel) root cortical parenchyma cells was delayed relative to the initiation of non-xylem cell formation by decreasing the temperature to 10°C. Once xylem or non-xylem cell formation was initiated, the respective production rates were not temperature sensitive within the 10-25°C range. Lowering the temperature after tracheary element differentiation had been initiated had no discernible effect on the differentiation process. Increasing the temperature from 10°C to 25°C on day seven did not give rise to an immediate and rapid production of tracheary elements. Shininger (1979a) postulated that xylogenesis in this system consists of at least two phases in regard to cold sensitivity: an early cold-sensitive phase detectable as soon as the cultures are started and a subsequent cold-insensitive phase that starts after xylogenesis has commenced.

The *in vitro* results with the inhibitory effects of low temperatures on xylem differentiation are puzzling when we consider that this developmental process occurs at much lower temperatures under natural conditions. For example, tracheary element differentiation has been shown to occur in *Pinus sylvestris* seedlings grown on a light/dark temperature cycle of 17.5/10°C (Denne, 1971).

High Temperature Effects

Relatively high temperatures stimulate xylogenesis, and Gautheret (1961) reported that explants of *Helianthus tuberosus* incubated at 31°C produced a woody type of growth. Similar results were reported by Phillips and Dodds (1977), and they found that changes in numbers of tracheary elements in response to different temperatures was paralleled by similar changes in the total cell number. In root explants of *Pisum*, on the other hand, cell proliferation rates were relatively less influenced by temperature changes (10-25°C) than tracheary element counts (Shininger, 1979a, b). According to Phillips and Dodds (1977) the optimum temperature for cell division in artichoke tuber explants was 30°C, whereas the highest proportion of tracheary elements to undifferentiated cells was found at 33°C. This finding was in agreement with the results of Gautheret (1961) and Naik (1965). Both of these workers found extremely high numbers of tracheary elements were formed in artichoke tissue cultured for prolonged periods at temperatures above 30°C. In fact, Naik (1965) found that 43 per cent of the cultured cells had differentiated into tracheary elements after 28 days of culture at 35°C. Although the reason for the great stimulation of xylogenesis at these elevated temperatures is unknown, Roberts (1976) suggested that either the rate of some metabolic pathway leading to xylogenesis has a high temperature optimum, or that the biosynthesis of some xylogenic hormone is stimulated by the higher temperatures (see Syōno and Furuya, 1971).

Water

The investigator of water relations in relation to xylogenesis *in vitro* is primarily interested in the effects of lower osmotic potential instead of waterlogged conditions. Cells obviously thrive during immersion in a liquid medium, provided that the system has been adjusted osmotically and receives sufficient agitation. Fluctuations in the water potential of a nutrient medium may influence xylogenesis indirectly by causing changes in the rate of mitotic activity. The rate of cell division, as reflected by DNA content, was greatly stimulated as the turgor pressure of the cells increased from five to six bars (Kirkham, Gardner and Gerloff, 1972). Another possibility concerns the induction of ethylene biosynthesis in the cultures subjected to water stress. Several studies have shown that leaves placed under water stress produce ethylene at increased rates (see Lieberman, 1979).

Doley and Leyton (1970) found that lowering the osmotic potential of the medium, with either sucrose or polyethylene glycol (PEG), stimulated xylogenesis in wound callus of *Fraxinus*. As the water potential became more negative the optimum concentration of IAA required for the induction of xylogenesis increased. The *Fraxinus* callus was unusual in that this system differentiated both sclereids and tracheary elements. Xylem differentiation, however, was invariably associated with the activity of a vascular cambium. Bornman and Huber (1979) examined the effects of water stress induced with PEG on xylogenesis in callus of *Nicotiana tabacum*. There were significantly more tracheary elements formed in callus stressed with 10 and 15 per cent PEG than observed in unstressed controls. At 10 per cent PEG a peak in phenylalanine ammonia-lyase activity coincided with maximal numbers of tracheary elements, whereas 15 per cent PEG was the optimal concentration for the biosynthesis of proline by the cultured callus. Numerous workers have reported the accumulation of proline resulting from osmotic stress (see Parsons, 1982). Earlier Roberts and Baba (1968) had observed that exogenous proline enhanced auxin-cytokinin-induced xylogenesis in *Coleus* internodal slices. A different view concerning PEG-induced stress was given by Wright and Northcote (1973). These workers were unsuccessful in inducing cytodifferentiation in sycamore callus explants cultured on media containing various levels of PEG. The variability of the results found with PEG may be traced to the presence of toxic contaminants in commercial preparations of the chemical (see Roberts, 1976).

Tracheary element formation in explants of lettuce (*Lactuca sativa* L. cv. Romana) pith was affected by the relative amount of surface water adhering to the explant, i.e. the residual water film on the lower side next to the medium (Johnson and Roberts, 1978). After surgical preparation and rinsing of the explants, the excess water was removed either by blotting with Whatman No. 1 filter paper or with Whatman GF/D glass microfibre filter circles. Counts of newly-formed tracheary elements were made following a seven-day incubation period on a xylogenic agar medium. Additional explants were rehydrated with microlitre amounts of water following blotting, and subsequently cultured in the same manner as the non-rehydrated. The rehydrated explants produced significantly greater numbers of tracheary elements in comparison to explants that had been blotted and not rehydrated (Johnson and Roberts, 1978). One explanation for these findings is that the uptake of nutrients and hormones from the agar medium is favoured by some optimum aqueous 'seal' between agar and tissue, and this optimum hydrature is indirectly shown by the fluctuations in the numbers of tracheary elements differentiated. Another possibility is that the water level controls ethylene biosynthesis, and this gaseous hormone, in turn, influences tracheary element differentiation. Waterlogging of tomato roots resulted in the biosynthesis and accumulation of 1-aminocyclopropane-1-carboxylic acid (ACC), which is the immediate precursor to ethylene (Bradford and Yang, 1980; Adams and Yang, 1981), and a similar phenomenon may occur in the waterlogged cells of the lettuce pith explant. The ACC thus formed, instead of being converted into ethylene and released as an atmospheric gas by the cells adjacent to the surface of the explant, would be available to synergize the auxin-cytokinin xylogenesis process occurring at more remote sites within the explant.

Light

The characteristics of radiation which affect cultured plant tissues are intensity, spectral quality, and the length of the daily exposure (Seibert and Kadkade, 1980). It is unfortunate that relatively little research has been reported on the effects of light on the *in vitro* differentiation of tracheary elements (see Roberts, 1976). Light is obviously not a necessity for the initiation of xylem differentiation, since numerous studies on cytodifferentiation have employed cultures maintained in total darkness throughout the experimental period. There is little agreement among research workers whether white (fluorescent) light stimulates or inhibits xylogenesis *in vitro*. Phillips and Dodds (1977) found that a brief exposure of *Helianthus tuberosus* tuber explants to dim white light during the culture procedure had an inhibitory effect on both cell division and tracheary element differentiation in comparison to the use of dim green light during culturing. Exposure to continuous light (4.8 lm/m^2, universal white, fluorescent) during the 72 hr culture period had no discernible effect on cell number but was inhibitory to xylogenesis by 40 per cent when compared to dark-grown controls (Phillips and Dodds, 1977). The inhibitory effects of light on cell division in cultured artichoke explants has been observed by other workers (Yeoman and Davidson, 1971; Davidson and Yeoman, 1974; James and Davidson, 1977).

A Japanese team of investigators found that light was a necessity for the induction of xylogenesis in slices of carrot root phloem cultured on a nutrient medium containing 2,4-D as the exogenous hormone. Explants cultured under the same cultural conditions in complete darkness never formed tracheary elements (Mizuno, Komamine and Shimokoriyama, 1971). Since the addition of cytokinin to the 2,4-D medium induced the differentiation of tracheary elements under both light and dark conditions, it was postulated that light initiated the biosynthesis of a xylogenic hormone. The cytokinin synthesised in the light by the carrot system was later isolated and identified (Mizuno and Komamine, 1978).

Does *in vitro* xylogenesis respond to the red and far-red wave lengths that serve to regulate phytochrome-mediated responses? Although this is still an open question, there is evidence to the affirmative. A greater number of tracheary elements were produced in the hypocotyl of *Sinapis alba* L. seedlings in the presence of the far-red absorbing form of phytochrome (P730) compared to dark-grown controls. In addition, it was found that the pattern of cytodifferentiation within the vascular bundles and the orientation of the bundles within the hypocotyl were the same in etiolated and illuminated plants (Kleiber and Mohr, 1967). The strongest evidence suggesting that phytochrome plays a role in the regulation of xylogenesis is found in the literature on the effects of light on the enzyme systems associated with lignification. For example, a causal chain of events can be traced between light-induced induction of PAL activity, lignin production, and xylogenesis. Light exerts an influence on all of the enzymes of the flavanoid pathway as shown in cell cultures of parsley (*Petroselinum hortense*), and these light-sensitive enzymes fall into two different groups based on their mode of action (Grisebach and Hahlbrock, 1974). The first group includes PAL, cinnamic acid 4-hydroxylase, and *p*-coumarate:CoA ligase. The second group is represented by flavonone synthetase, glucosyltransferase, apiosyltransferase, UDP-apiose synthetase, chalcone flavonone isomerase, and others. The first group of enzymes catalyses the production of phenylpropanoid compounds, whereas the second group are involved in flavone glycoside biosynthesis. The activities of all enzymes of both groups were increased after exposure to light (Grisebach and Hahlbrock, 1974). Experiments involving transcription and translation inhibitors have shown that light-induced increases in PAL activity are due to *de novo* synthesis and not to light activation of a preexisting inactive form of the enzyme (Wellmann and Schoffer, 1975). Haddon and Northcote (1975) found that increased lignin biosynthesis was correlated with the appearance of PAL activity in *Phaseolus* callus, and additional observations on this phenomenon were given by Dudley and Northcote (1979). PAL was designated as a marker enzyme for tracheary element differentiation by Rubery and Fosket (1969). PAL activity followed, over the same time course, xylem differentiation in cultured *Coleus* internodes. Both PAL activity and tracheary element differentiation peaked after four days of culture, and both rapidly declined thereafter. These same investigators found that conditions which favoured xylem differentiation in soybean (*Glycine max* L. cv. Biloxi) callus cultures also enhanced PAL activity. Haddon and Northcote (1976) also found that PAL activity increased during cytodifferentiation in callus of bean (*Phaseolus vulgaris* L.). Additional information on the role of light in the regulation of enzyme levels in the phenylpropanoid pathway can be found in the publication by Lamb (1979) and the review by Hahlbrock and Grisebach (1979).

Gas Composition

The gases normally encountered during a plant tissue culture experiment include oxygen, carbon dioxide, ethylene and ozone. Aside from ethylene little interest has been shown in the possible role(s) of gases in tracheary element differentiation.

Oxygen

The level of dissolved oxygen in suspension cultures of carrot tissue apparently has a regulatory effect on differentiation in this system. Below a critical concentration of oxygen embryogenesis was initiated, whereas higher concentrations of the gas favoured rhizogenesis (Kessell and Carr, 1972). The possible effect of anaerobiosis on xylogenesis has not been examined, although this condition could indirectly modify the synergistic effect of ethylene on the process. Lowering the level of oxygen to 5 per cent was shown to reduce ethylene-induced growth retardation, inhibit ethylene incorporation into the tissue and suppress the oxidation of ethylene to carbon dioxide (Beyer, 1979).

Carbon Dioxide

According to Martin (1980) there is little evidence in the literature that exogenous carbon dioxide plays a significant role in the growth of plant cell cultures in the presence of other carbon sources. There is, however, the possibility that carbon dioxide influences xylogenesis indirectly via ethylene (see Roberts, 1976). Carbon dioxide stimulated ethylene biosynthesis in tobacco leaf discs (Aharoni and Lieberman, 1979) and in intact sunflower plants (Bassi and Spencer, 1982). Other workers found that carbon dioxide stimulated xylogenesis in cultured peach mesocarp tissue (Bradley and Dahmen, 1971). Although there is no evidence to date that carbon dioxide *per se* has any direct effect on xylem differentiation, a critical study should be undertaken on the induction of cytodifferentiation *in vitro* in the presence of various combinations of carbon dioxide and ethylene.

Ethylene

Very low concentrations of ethylene have a stimulatory effect on the *in vitro* induction of xylogenesis in the presence of auxin and cytokinin (Roberts and Baba, 1978; Miller and Roberts, 1982), and ethylene may be a critical hormone in the initiation of tracheary element differentiation (Roberts *et al.*, 1982). Details concerning the relationships between ethylene and xylogenesis will be considered in this chapter under mechanical stress (see also Chapter 6).

Ozone

Ozone had an inhibitory effect on tracheary element formation in wounded stem internodes of *Coleus* and cultured callus of *Parthenocissus* (Rier, 1976). Precautions should be taken during the culture procedure, since toxic levels of ozone may be present in an enclosed transfer chamber as a result of prolonged ultraviolet irradiation emitted from a germicidal lamp.

Mechanical Stress

Plant tissues rapidly synthesise ethylene in response to external pressure or tension, vibration, wind, or touch (Jaffe, 1980). The best known thigmomorphogenic growth responses of ethylene include epinasty, decreased shoot elongation, increased radial enlargement of cells and diageotropism. Less appreciated are changes in mitotic activity and cytodifferentiation that can be attributed to stress-induced ethylene biosynthesis. The role of pressure in cell differentiation was first demonstrated by Brown and Sax (1962). Longitudinal bark strips of *Populus trichocarpa* and *Pinus strobus*, separated from the bole wood during early spring, were encased in plastic to prevent desiccation. The bark strips formed an extensive undifferentiated callus along the cambial zone. Similar bark strips held firmly against the tree with external pressure (0.25 to 1.0 atm) produced normal xylem and phloem elements from cambial derivatives. In a subsequent paper Brown (1964) described the effects of external pressure on the induction of tracheary element differentiation in explants prepared from *Populus deltoides*. Other workers have reported that the application of mechanical stress resulted in an increased rate of cell proliferation and changes in the plane of cell division under *in vitro* (Yeoman and Brown, 1971) and *in vivo* (Lintilhac and Vesecky, 1981) conditions.

Tissue sampling has demonstrated that ethylene is produced by the secondary xylem during the application of physical stress. Mechanically-stressed branches of *Pinus strobus*, *Pyrus malus*, and *Prunus persica* trees produced more secondary xylem than non-stressed control branches (Leopold, 1972; Brown and Leopold, 1973). The bending stress of wind on plants is considered to stimulate ethylene formation which results in increased radial growth (Jacobs, 1954; Larson, 1965; Hunt and Jaffe, 1980). The bending of apple seedlings in order to place the shoot apex in a horizontal position caused increased ethylene levels in the stressed shoots (Robitaille and Leopold, 1974). Ethylene concentrations were found to be higher on the under side, compared to the upper side, of the horizontally positioned apple shoots (Robitaille, 1975).

Some responses of the secondary xylem to mechanical stress include greater numbers of cells, smaller cells, thicker secondary cell walls, and increased lignification. Although compression and shaking resulted in thigmomorphogenic responses in *Pinus resinosa*, twisting or torque stress produced the greatest effect (Quirk, Smith and Freese, 1976). Mechanically-stressed beans, sunflowers, and liquidambar trees produced more secondary xylem and thicker xylem cells than non-stressed plants (Jaffe, 1980). Plants of *Passiflora caerulea* form tendrils that exhibit thigmomorphogenic responses, and contact stimulus-induced lignification of the xylem elements within the tendrils (Reinhold, Sachs and Vislovska, 1972).

The application of either ethylene gas or (2-chloroethyl) phosphonic acid (ethephon) to a plant alters the morphology and quantity of secondary xylem. Ethylene is the only plant hormone, when given exogenously, that precisely mimics all of the thigmomorphogenic responses resulting from mechanical perturbation (Jaffe and Biro, 1979). The application of ethephon to shoots of pine (Brown and Leopold, 1973) and apple (Robitaille and Leopold, 1974) produced localised increases in stem diameter.

The hormonal physiology of reaction wood formation is not clearly understood at this time, and both auxin and ethylene may be involved in this differentiation process. Deviation from the vertical induces the formation of tension wood in angiosperms and compression wood in gymnosperms (see Jane, 1970; Timell, 1973). Briefly, the tension wood of angiosperms contains vessels with reduced width and in lesser numbers, and the fibres have a thick gelatinous layer interior to the S layers (Esau, 1977). Compression wood of gymnosperms is characterised by the presence of short tracheids with cell walls appearing rounded in transections. In addition, the tracheids are strongly lignified and usually the S3 layer of the secondary wall is lacking. Experimental evidence suggests a relationship exists between stress-induced ethylene biosynthesis and reaction wood formation. Nelson and Hillis (1978) performed experiments on the induction of tension wood formation in *Eucalyptus gomphocephala* by placing the seedlings in a horizontal position. The stressed plants had 60-80 per cent tension wood by volume in the upper halves and 0-10 per cent in the lower halves. The upper halves had greater amounts of ethylene than the lower halves and vertical seedling halves (Nelson and Hillis, 1978).

Compression wood has been induced by 1-N-naphthylphthalamic acid, an IAA transport inhibitor (Yamaguchi, Itoh and Shimaji, 1980). The latter workers suggested that compression wood formation may be a general characteristic of agents effective in blocking auxin transport. Ethylene biosynthesis induced by mechanical perturbation has been reported to block the polar transport of auxin (Mitchell, 1977).

Acidity

In general, most plant tissue culture media are buffered prior to autoclaving within the range of pH 5-6. Media vary considerably in regard to their buffering capacity, and, in some cases, the pH may shift dramatically during a tissue culture experiment (Martin, 1980). Virtually no work has been done on the possible effects of pH on the cytodifferentiation of tracheary elements. A preliminary study by Datta and colleagues (1975) showed that the initial pH of a medium apparently had some effects on secondary xylem production in explants of *Plumeria*. A combination of pH and autoclaving may be detrimental to a culture because of the instability of some of the components of the medium to pH, heat, and pressure. For example, thiamine-HCl is rapidly destroyed on heating at pH values above 5.5 (Windholz, 1976). Some type of control over fluctuations in pH during the period of a xylogenesis experiment may be advantageous. Shiraishi (personal communication) conducted experiments on xylem differentiation in lettuce pith explants cultured on a filter paper platform (Dodds and Roberts, 1982) with a liquid Murashige and Skoog (1962) induction medium. The initial medium, adjusted to pH 5.5 prior to autoclaving, registered approximately 5.1 after autoclaving. After 7 days of culture the pH of the medium had risen to 6.8-6.9. Numbers of tracheary elements formed were significantly increased by transferring the explants, after 4 days of culture, to another culture tube containing a fresh medium buffered to the original level of 5.5. Since other factors were obviously involved, i.e. replenishment of the induction medium, the relative importance of pH in these experiments is still unknown.

References

Adams, D.O. and Yang, S.F. (1981) 'Ethylene, the gaseous plant hormone: mechanism and regulation of biosynthesis', *Trends in Biochem. Sci.*, *6*, 161-3

Aharoni, N. and Lieberman, M. (1979) 'Ethylene as a regulator of senescence in tobacco leaf discs', *Plant Physiol.*, *64*, 801-4

Bassi, P.K. and Spencer, M.S. (1982) 'Effect of carbon dioxide and light on ethylene production in intact sunflower plants', *Plant Physiol.*, *69*, 1222-5

Beyer, E.M., Jr. (1979) 'Effect of silver ion, carbon dioxide, and oxygen on ethylene action and metabolism', *Plant Physiol.*, *63*, 169-73

Bornman, C.H. and Huber, W. (1979) '*Nicotiana tabacum* callus studies. 9. Development in stressed explants', *Biochemie Physiol. Pflanzen*, *174*, 345-56

Bradford, K.J. and Yang, S.F. (1980) 'Xylem transport of 1-aminocyclopropane-1-carboxylic acid, an ethylene precursor, in waterlogged tomato plants', *Plant Physiol.*, *65*, 322-6

Bradley, M.V. and Dahmen, W.J. (1971) 'Cytohistological effects of ethylene, 2,4-D, kinetin and carbon dioxide on peach mesocarp callus cultured *in vitro*', *Phytomorphology*, *21*, 154-64

Brown, C.L. (1964) 'The influence of external pressure on the differentiation of cells and tissues cultured *in vitro*' in M.H. Zimmermann (ed.), *The Formation of Wood in Forest Trees*, Academic Press, New York, pp. 389-404

Brown, C.L. and Sax, K. (1962) 'The influence of pressure on the differentiation of secondary tissues', *Am. J. Bot.*, *49*, 683-91

Brown, K.M. and Leopold, A.C. (1973) 'Ethylene and the regulation of growth in pine', *Can. J. For. Res.*, *3*, 143-5

Chafe, S.C. and Durzan, D.J. (1973) 'Tannin inclusions in cell suspension cultures of white spruce', *Planta*, *113*, 251-62

Datta, S.K., Chakrabarti, K. and Datta, P.C. (1975) '*In vitro* effect of acidity levels on xylem differentiation in *Plumeria*', *Current Sci.*, *44*, 814-16

Davidson, A.W. and Yeoman, M.M. (1974) 'A phytochrome-mediated sequence of reactions regulating cell division in developing callus cultures', *Ann. Bot.*, *38*, 545-54

Denne, M.B. (1971) 'Temperature and tracheid development in *Pinus sylvestris* seedlings', *J. Exp. Bot.*, *22*, 362-70

Denne, M.B. and Dodd, R.S. (1981) 'The environmental control of xylem differentiation' in J.R. Barnett (ed.), *Xylem Cell Development*, Castlehouse Publications Ltd., Tunbridge Wells, U.K., pp. 236-55

Dodds, J.H. and Roberts, L.W. (1982) *Experiments in Plant Tissue Culture*, University Press, Cambridge

Doley, D. and Leyton, L. (1970) 'Effects of growth regulating substances and water potential on the development of wound callus in *Fraxinus*', *New Phytol.*, *69*, 87-102

Dudley, K. and Northcote, D.H. (1979) 'Regulation of induction of phenylalanine ammonia-lyase in suspension cultures of *Phaseolus vulgaris*', *Planta*, *146*, 433-40

Durzan, D.J., Chafe, S.C. and Lopushanski, S.M. (1973) 'Effects of environmental changes on sugars, tannins, and organized growth in cell suspension cultures of white spruce', *Planta*, *133*, 241-9

Esau, K. (1977) *Anatomy of Seed Plants*, 2nd edn., J. Wiley and Sons, New York.

Gautheret, R.J. (1961) 'Action de la lumière et de la température sur la néoformation de racines par des tissus de Topinambour cultivés *in vitro*', *Compte. Rendu. Acad. Sci.*, *250*, 2791-6

Gautheret, R.J. (1969) 'Investigations on the root formation in the tissues of *Helianthus tuberosus* cultured *in vitro*', *Am. J. Bot.*, *56*, 702-17

Grisebach, H. and Hahlbrock, K. (1974) 'Enzymology and regulation of flavanoid and lignin biosynthesis in plants and plant cell suspension cultures' in V.C. Runeckles and E.E. Conn (eds.), *Metabolism and Regulation of Secondary Plant Products*, Academic Press, New York, pp. 22-52

Grout, B.W.W. and Aston, M.J. (1977) 'Transplanting of cauliflower plants regenerated from meristem culture. I. Water loss and water transfer related to changes in leaf wax and to xylem regeneration', *Hort. Res.*, *17*, 1-7

Haddon, L.E. and Northcote, D.H. (1975) 'Quantitative measurement of bean callus differentiation', *J. Cell Sci.*, *17*, 11-26

Haddon, L.E. and Northcote, D.H. (1976) 'Correlation of the induction of various enzymes concerned with phenylpropanoid and lignin synthesis during differentiation of bean callus (*Phaseolus vulgaris* L.)', *Planta*, *128*, 255-62

Hahlbrock, K. and Grisebach, H. (1979) 'Enzymic controls in the biosynthesis of lignin and flavonoids', *Ann. Rev. Plant Physiol.*, *30*, 105-30

Hunt, E.R., Jr. and Jaffe, M.J. (1980) 'Thigmomorphogenesis: the interaction of wind and temperature in the field on the growth of *Phaseolus vulgaris* L.', *Ann. Bot.*, *45*, 665-72

Jacobs, M.R. (1954) 'The effect of wind sway on the form and development of *Pinus radiata* D. Don', *Aust. J. Bot.*, *2*, 35-51

Jacquoit, C. (1947) 'Effet inhibiteur des tannins sur le développement des cultures *in vitro* du cambium de certaines arbres forestiers', *Compte. Rendu. Acad. Sci.*, *225*, 434-6

Jaffe, M.J. (1980) 'Morphogenetic responses of plants to mechanical stimuli or stress', *BioSci.*, *30*, 239-43

Jaffe, M.J. and Biro, R. (1979) 'Thigmomorphogenesis: the effect of mechanical perturbation on the growth of plants, with special reference to anatomical changes, the role of ethylene, and interaction with other environmental stresses' in H. Mussell and R.C. Staples (eds.), *Stress Physiology in Crop Plants*, J. Wiley and Sons, New York, pp. 25-69

James, D.J. and Davidson, A.W. (1977) 'Phytochrome control of phenylalanine ammonia-lyase levels and the regulation of cell division in artichoke callus cultures', *Ann. Bot.*, *41*, 873-7

Jane, F.W. (1970) *The Structure of Wood*, 2nd edn, Adam and Charles Black, London

Johnson, D.C. and Roberts, L.W. (1978) 'Water stress influences xylogenesis in cultured explants of *Lactuca*', *Phytomorphology*, *28*, 207-9

Kessell, R.H.J. and Carr, A.H. (1972) 'The effect of dissolved oxygen concentration on growth and differentiation of carrot (*Daucus carota*) tissue', *J. Exp. Bot.*, *23*, 996-1007

Kirkham, M.B., Gardner, W.R. and Gerloff, G.C. (1972) 'Regulation of cell division and cell enlargement by turgor pressure', *Plant Physiol.*, *49*, 961-2

Kleiber, H. and Mohr, H. (1967) 'Vom Einfluss des Phytochroms auf die Xylemdifferzierung im Hypokotyl des Senfkimlings (*Sinapis alba* L.)', *Planta*, *76*, 85-92

Lamb, C.J. (1979) 'Regulation of enzyme levels in phenylpropanoid biosynthesis: characterization of the modulation by light and pathway intermediates', *Arch. Biochem. Biophys.*, *192*, 311-17

Larson, P.R. (1965) 'Stem form of young *Larix* as influenced by wind and pruning', *Forest Sci.*, *11*, 412-24

Leopold, A.C. (1972) 'Ethylene as a plant hormone' in H. Kaldewey and Y. Vardar (eds.), *Hormonal Regulation in Plant Growth and Development*, Verlag Chemie, Weinheim, pp. 245-62

Lieberman, M. (1979) 'Biosynthesis and action of ethylene', *Ann. Rev. Plant Physiol.*, *30*, 533-91

Lintilhac, P.M. and Vesecky, T.B. (1981) 'Mechanical stress and cell wall orientation in plants. II. The application of controlled directional stress to growing plants; with a discussion on the nature of the wound reaction', *Am. J. Bot.*, *68*, 1222-30

McDaniel, R.G. (1982) 'The physiology of temperature effects on plants' in M.N. Christiansen and C.F. Lewis (eds.), *Breeding Plants for Less Favorable Environments*, J. Wiley and Sons, New York, pp. 13-45

Martin, S.M. (1980) 'Environmental factors. B. Temperature, aeration, and pH' in E.J. Staba (ed.), *Plant Tissue Culture as a Source of Biochemicals*, CRC Press, Boca Raton, Florida, pp. 143-8

Miller, A.R. and Roberts, L.W. (1982) 'Regulation of tracheary element differentiation by exogenous L-methionine in callus of soybean cultivars', *Ann. Bot.* (in press)

Mitchell, C.A. (1977) 'Influences of mechanical stress on auxin-stimulated growth of excised pea stem sections', *Physiol. Plant*, *41*, 129-34

Mizuno, K. and Komamine, A. (1978) 'Isolation and identification of substances inducing formation of tracheary elements in cultured carrot-root slices', *Planta*, *138*, 59-62

Mizuno, K., Komamine, A. and Shimokoriyama, M. (1971) 'Vessel element formation in cultured carrot-root phloem', *Plant Cell Physiol.*, *12*, 823-30

Murashige, T. (1974) 'Plant propagation through tissue cultures', *Ann. Rev. Plant Physiol.*, *25*, 135-66

Murashige, T. and Skoog, F. (1962) 'A revised medium for rapid growth and bioassays with tobacco tissue cultures', *Physiol. Plant.*, *15*, 473-97

Naik, G.G. (1965) 'Studies on the effects of temperature on the growth of plant tissue cultures', M.Sc. Thesis, University of Edinburgh, Scotland.

Nelson, N.D. and Hillis, W.E. (1978) 'Ethylene and tension wood formation in *Eucalyptus gomphocephala*', *Wood Sci. Tech.*, *12*, 309-15

Parsons, L.R. (1982) 'Plant responses to water stress' in M.N. Christiansen and C.F. Lewis (eds.), *Breeding Plants for Less Favorable Environments*, J. Wiley and Sons, New York, pp. 175-92

Phillips, R. and Dodds, J.H. (1977) 'Rapid differentiation of tracheary elements in cultured explants of Jerusalem artichoke', *Planta*, *135*, 207-12

Quirk, J.T., Smith, D.M. and Freese, F. (1975) 'Effect of mechanical stress on growth and anatomical structure of red pine (*Pinus resinosa* Ait.): torque stress', *Can. J. For. Res.*, *5*, 691-9

Reinhold, L., Sachs, T. and Vislovska, L. (1972) 'The role of auxin in thigmotropism' in D.J. Carr (ed.), *Plant Growth Substances, 1970*, Springer-Verlag, Berlin, pp. 731-7

Rier, J.P., Jr. (1976) 'Ozone and vascular tissue differentiation in plants', U.S. Environmental Protection Agency, Washington, D.C. (EPA-600/3-76-068)

Roberts, L.W. (1976) *Cytodifferentiation in Plants: Xylogenesis as a Model System*, University Press, Cambridge

Roberts, L.W. and Baba, S. (1968) 'Effect of proline on wound vessel member formation', *Plant Cell Physiol.*, *9*, 353-60

Roberts, L.W. and Baba, S. (1978) 'Exogenous methionine as a nutrient supplement for the induction of xylogenesis in lettuce pith explants', *Ann. Bot.*, *42*, 375-9

Roberts, L.W., Baba, S., Shiraishi, T. and Miller, A.R. (1982) 'Progress in cytodifferentiation under *in vitro* conditions', *Proc. V Internat. Congress of Plant Tissue and Cell Culture*, IAPTC, Tokyo

Robitaille, H.A. (1975) 'Stress ethylene production in apple shoots', *J. Am. Soc. Hort. Sci.*, *100*, 524-7

Robitaille, H.A. and Leopold, A.C. (1974) 'Ethylene and the regulation of apple stem growth under stress', *Physiol. Plant*, *32*, 301-4

Rubery, P.H. and Fosket, D.E. (1969) 'Changes in phenylalanine ammonia-lyase during xylem differentiation in *Coleus* and soybean', *Planta*, *87*, 54-62

Savidge, R.A. and Wareing, P.F. (1981) 'Plant-growth regulators and the differentiation of vascular elements' in J.R. Barnett (ed.), *Xylem Cell Development*, Castlehouse Publications Ltd., Tunbridge Wells, U.K., pp. 192-235

Seibert, M. and Kadkade, P.G. (1980) 'Environmental factors. A. Light' in E.J. Staba (ed.), *Plant Tissue Culture as a Source of Biochemicals*, CRC Press, Boca Raton, Florida, pp. 123-41

Shininger, T.L. (1979a) 'Xylem and nonxylem cell formation in cytokinin-stimulated root tissue. Quantitative analysis of temperature effects', *Proc. Nat. Acad. Sci.* (USA), *76*, 1921-3

Shininger, T.L. (1979b) 'The control of vascular development', *Ann. Rev. Plant Physiol.*, *30*, 313-37

Syōno, K. and Furuya, T. (1971) 'Effects of temperature on the cytokinin requirement of tobacco calluses', *Plant Cell Physiol.*, *12*, 61-71

Timell, T.E. (1973) 'Ultrastructure of the dormant and active cambial zones and the dormant phloem associated with formation of normal and compression woods in *Picea abies* (L.) Karst', Pub. 96, Office of Public Service and Continuing Education, State University of New York College of Environmental Science and Forestry, Syracuse

Wellmann, E. and Schoffer, P. (1975) 'Phytochrome-mediated *de novo* synthesis of phenylalanine ammonia-lyase in cell suspension cultures of parsley', *Plant Physiol.*, *55*, 822-8

Windholz, M. (ed.) (1976) *The Merck Index. An Encyclopedia of Chemicals and Drugs*, 9th edn, Merck, Rahway, New Jersey, USA

Wright, K. and Northcote, D.H. (1973) 'Differences in ploidy and degree of intercellular contact in differentiating and non-differentiating sycamore calluses', *J. Cell Sci.*, *12*, 37-53

Yamaguchi, K., Itoh, T. and Shimaji, K. (1980) 'Compression wood induced by 1-N-naphthylphthalamic acid (NPA), an IAA transport inhibitor', *Wood Sci. Tech.*, *14*, 181-5

Yeoman, M.M. and Brown, R. (1971) 'Effects of mechanical stress on the plane of cell division in developing callus cultures', *Ann. Bot.*, *35*, 1101-12

Yeoman, M.M. and Davidson, A.W. (1971) 'Effect of light on cell division in developing callus cultures', *Ann. Bot.*, *35*, 1085-1100

9 THE USE OF PROTOPLAST TECHNOLOGY IN TISSUE CULTURE OF TREES

John H. Dodds

History

Isolated plant protoplasts can be described as naked plant cells. Protoplasts are plant cells where the cell wall has been removed either by a mechanical or enzymic method. The original way of preparing protoplasts was by plasmolysis of the tissue followed by mechanical abrasion. The cell wall would then allow release of the plasmolysed protoplast. The mechanical method of protoplast isolation has the disadvantages that it is very tedious and time consuming and more importantly only very small numbers of protoplasts can be isolated. One of the advantages, however, of mechanical isolation is that there are no enzyme or enzyme impurities present to attack and weaken the plasma membrane of the isolated protoplast. In the 1960s techniques were developed for the isolation of protoplasts from leaf tissue using the enzymes cellulase and pectinase: these enzymic techniques offer a quick and efficient method for the isolation of very large numbers of protoplasts.

General Methods for Protoplast Isolation

A schematic representation of the steps used in protoplast isolation is shown in Figure 9.1. In early studies protoplasts were normally isolated from leaf mesophyll cells, however similar techniques are now applied to stem, callus, and root tissues. The material is first surface-sterilised in a 10 per cent Clorox solution then rinsed in sterile distilled water several times to remove excess hypochlorite. In some tissues it is advantageous to pre-plasmolyse the tissue before treatment with the enzyme mixture. Plant cell walls consist of a complex mixture of cellulose, hemicellulose, pectins, proteins and lipids and it is necessary therefore to treat the wall with a mixture of enzymes to facilitate total wall digestion. The normal enzyme mixture used is Cellulase and Macerozyme but a range of other enzymes are now available.

Once the protoplasts have been liberated into the surrounding medium it is necessary to remove the enzyme mixture, together with cell

Figure 9.1: The Leaf is Surface Sterilised, Rinsed Repeatedly in Distilled Water, and the Cells are Plasmolysed in a Mannitol Solution. The lower epidermal layer is removed to aid enzyme penetration. Following treatment with one or more wall degrading enzymes a crude suspension of protoplasts is obtained

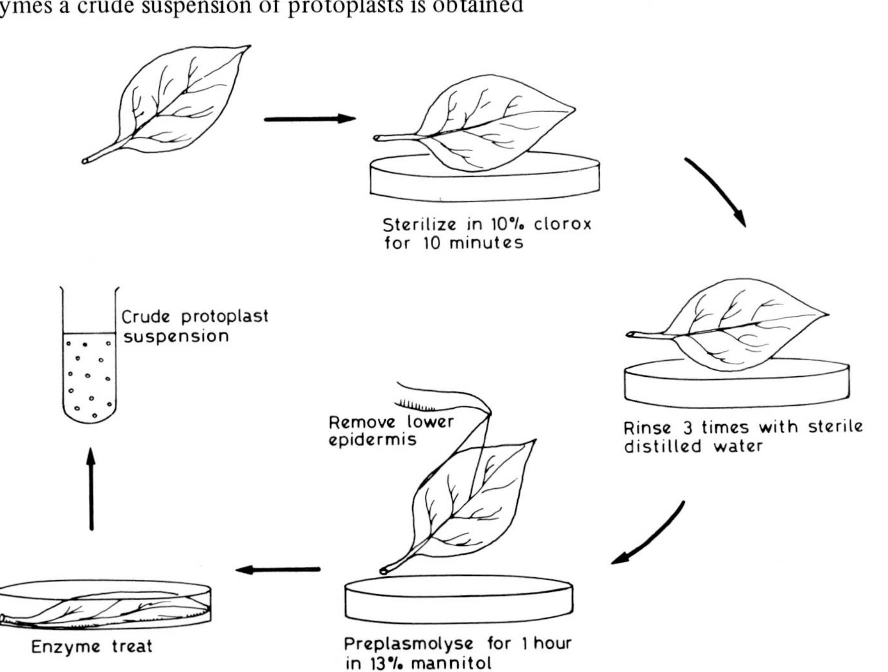

debris, which is done using the methods outlined in Figure 9.2. After filtration through a 45 μm mesh seive the protoplasts can be pelleted by gentle centrifugation and the pellet washed with fresh medium to remove enzyme mixture and cell debris. The result should be a clean preparation of isolated protoplasts similar to those in Figure 9.3.

Figure 9.2: The Crude Protoplast Suspension is Filtered Through a Nylon Mesh (45 μm) and the Filtrate is Centrifuged at 75 x g for 5 min. The protoplasts pellet is resuspended in 10 cm^3 of fresh culture medium and recentrifuged. Before transfer to culture media the preparation is examined and viability is determined

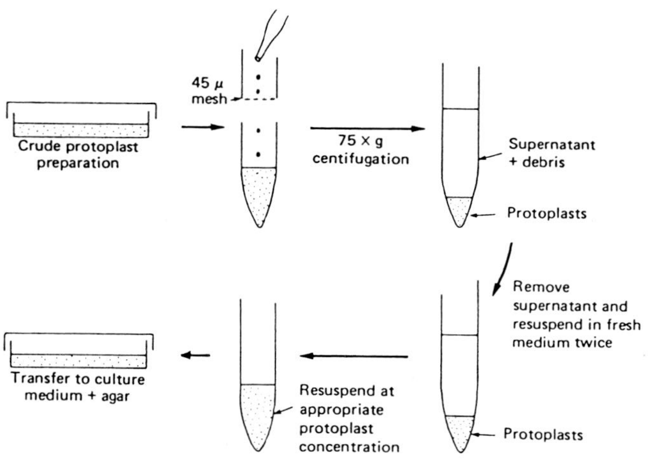

Figure 9.3: Isolated Plant Protoplast

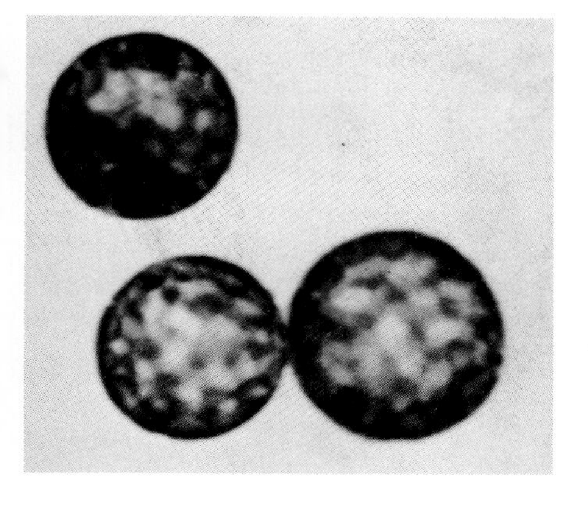

Culture of Isolated Protoplasts

A number of techniques have been devised for the culture of isolated protoplasts. Often the technique employed depends on the overall aim of the experiment. For example, it may be desirable to screen large numbers of protoplasts for the selection of single or small numbers of mutants, or it could be that one wants to study the development of a single cell. Nagata and Takebe (1971) developed a method for embedding protoplasts in a soft agar (1 per cent) matrix; this method gave some physical support to the protoplasts and ensured that they did not move, allowing for close observation of their development. Vasil and Vasil (1973) devised microculture chambers for protoplasts in a liquid medium, and hanging drops have also been used (Gleba, 1978). A technique for the culture of individual protoplasts in hanging drops known as the multi-drop array (MDA) method (Potrykus, Harms and Lörz, 1979) has allowed the screening of a very wide range of media on protoplast development.

Once cultured in an appropriate medium the protoplasts regenerate a new cell wall (Willison and Cocking, 1972) and undergo repeated mitotic divisions to form a callus (Lörz, Potrykus and Thomas, 1977). When the callus of some species is transferred to a medium lacking mannitol and auxin, embryo formation begins in the callus (Kameya and Uchimiya, 1972; Lörz *et al.*, 1977; Lörz and Potrykus, 1979). These embryos can be cultured and eventually develop into mature plants. Thus in some species it is possible to regenerate a whole plant from a single isolated protoplast, for example with many Solanaceous plants.

Uses of Plant Protoplasts and Possible Applications to Trees

The removal of the wall in the preparation of protoplasts opens up a wide range of possibilities for the *in vitro* manipulation of plant cells.

Use of Single Cells for Selection Systems

It is highly desirable to select plants that are suited to a particular environment or subject to a particular stress. For example some areas have soils that have a high salt (NaCl) concentration or have high levels of heavy metals (e.g. aluminium, lead) present. The screening of plants for resistance to these factors by conventional methods is very laborious and has a very high unit cost. The use of single cells (protoplasts) allows the screening of millions of individuals in a very short space of time.

Plants can then be regenerated from the resistant cells and tested for resistance at the whole plant and progeny levels. It may be possible to apply these techniques to cells from trees to select for, say, tolerance to heavy metals. The selected trees could then be used for reforestation of areas with high heavy metal content (e.g. mining waste tips).

Protoplast Fusion

As stated previously, when protoplasts are prepared, the boundary between the living cytoplasm and the external environment is the plasma membrane. The removal of the outer cell wall thus allows the plasma membranes of different protoplasts to come into contact with each other. The inclusion of a fusagenic agent such as polyethylene glycol in the medium allows protoplasts to fuse together in a similar way to two soap bubbles (Cocking, 1977; Constabel and Kao, 1974). Figure 9.4 shows, diagrammatically, the fusion of different protoplasts.

Figure 9.4: Fusion of Two Dissimilar Protoplasts (Denoted 1, 2) to Form a 'Hybrid'. A true hybrid is not formed until the nuclei fuse

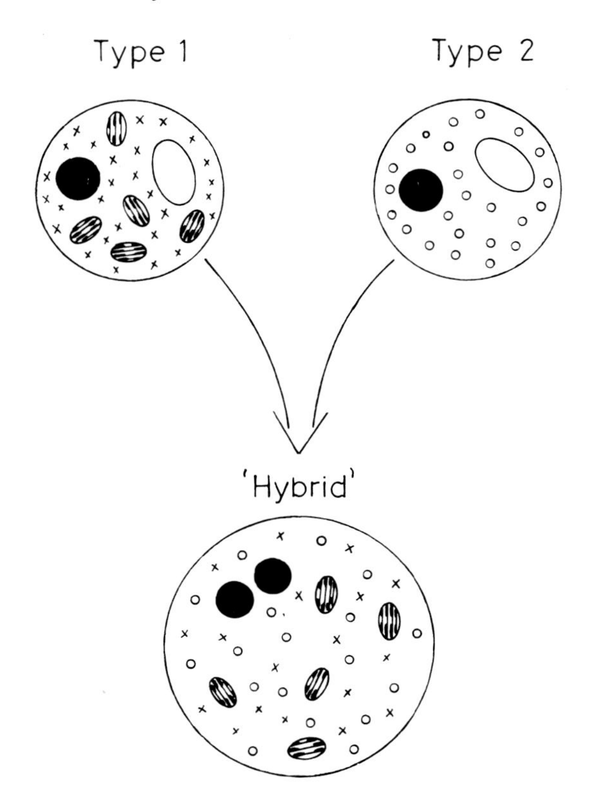

The development of the techniques of protoplast fusion have advanced very rapidly over the last decade (Ahkong *et al.*, 1975; Carlson, Smith and Dearing, 1972; Kao and Michayluk, 1974; Mastrangelo, 1979; Melchers and Labib, 1974; Power, Cummins and Cocking, 1970; Smith, Kao and Combatti, 1976). One of the key problems with protoplast fusion experiments is the selection of the desired fusion product. As shown in Figure 9.5, a number of fusion products can be formed, the difficulty is to devise methods for selection of the A.B. hybrid. A number of methods have been devised and are well discussed in a book by Chaleff (1981). The normal techniques are those of genetic complementation (Carlson, 1972), or autrotrophy (Constabel, 1978). Other techniques include visual selection and micromanipulation (Patnack *et al.*, 1982) and fluorescence cell sorting (Redenbaugh *et al.*, 1982).

Figure 9.5: The Fusion of Protoplasts A and B Results in a Binucleate Heterokaryon Containing the Cytoplasmic Contents of Both A and B. Fusion of the two nuclei results in a tetraploid hybrid or synkaryote. If one of the nuclei degenerates a cytoplasmic hybrid, heteroplast or cybrid is formed

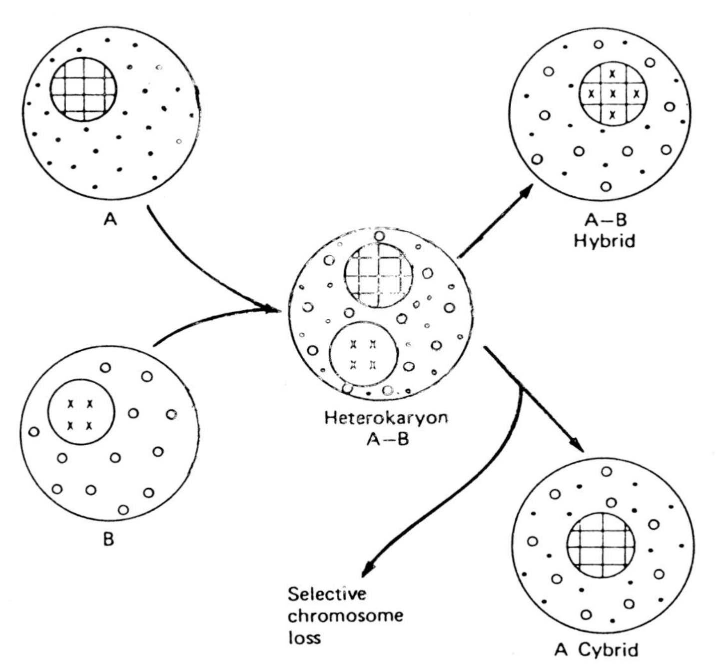

The fusion of the membranes of the two protoplasts in no way guarantees that a true somatic hybrid will be formed as a number of possible products exist. After fusion of the two plasma membranes the cytoplasm of the two cells becomes mixed, but the two nuclei have several possible fates: they may fuse together and form a true somatic hybrid, or there may be loss of some or all of the genetic information from one or both nuclei. If a nucleus is totally lost (or is purposely destroyed by microbeam irradiation) then a protoplast with hybrid cytoplasm is formed and this is known as a cybrid (cytoplasmic hybrid; Figure 9.5). The formation of hybrids between different crop plants with the idea of forming novel plants with many desirable characters from both parents has received much attention in the review literature (Cocking, 1977; Power and Cocking, 1971). Indeed, the experiments behind these ideas are now becoming a reality (Melchers, 1982).

The uses of protoplast fusion technology to tissue culture of trees are of great potential. With fruit trees many hybrids have been made by conventional sexual crosses that have very important economic potential. It is possible that protoplasts can be isolated and fused in a number of fruit trees to make 'novel' hybrids between fruits that are normally sexually incompatible.

The isolation and culture of protoplasts from timber crop trees is an area that is still in its infancy (David and David, 1979; Kirkby and Cheng, 1979; Winston, Parham and Kawstinen, 1975; David, David and Mateille, 1982). However, again the possibilities of novel crosses are important in the long-term breeding programmes employed in tree research.

In the field of protoplast culture very little work has been done with trees, compared to the enormous literature available on the *Solanaceae* family and the major economic food crops. It should be remembered that trees are of enormous economic importance both in terms of food production (fruits) and raw material production (wood and paper). More research is needed in the basic areas of study, such as conditions for isolation and culture of protoplasts and regeneration of plants from protoplast-derived callus, before more refined techniques can be developed for formation of non-sexual hybrids.

Genetic Manipulation

The removal of the cell wall in protoplast preparation, as well as allowing protoplasts to fuse together, also allows endocytosis by plant cells. The presence of the cell wall normally restricts the ability of a plant cell to take up foreign material. Once the wall is removed a range of materials can be incorporated into the cell. The materials incorporated into plant cells include chloroplasts, mitochondria, DNA, plasmids, bacteria, viruses and polystyrene beads (Davey and Cocking, 1972; Carlson, 1973; Cocking, 1977b).

The presence of these foreign materials in the protoplasts has normally been defined on the basis of microscopical analysis. It has become clear in recent years, however (Mastrangelo, 1979), that most of these materials are broken down by enzymes within the plant cell, i.e. destruction of foreign material in an almost immunological sense.

Attempts to incorporate foreign genetic material into plant cells have found a vehicle for transfer of genes thanks to studies of plant tumours (Braun, 1954, 1974). The bacterium that causes crown gall tumours, *Agrobacterium tumefaciens*, contains a plasmid which becomes incorporated in the plant nuclear DNA. This plasmid can be cleaved open and new genes inserted, using the plasmid as a vehicle to carry them to the plant nucleus (Ream and Gordon, 1982). Once the new genes have been inserted into the nucleus they will segregate in a normal Mendelian fashion. The techniques of plant genetic manipulation are still in their infancy but the potential of these techniques for agricultural improvement is enormous. These techniques may be applicable to trees to improve photosynthetic efficiency or to modify storage products. In the far distant future it may be possible to use these techniques to insert genes for nitrogen fixation into non-legume trees. Nevertheless, before these techniques become applicable a large amount of background work is required on the isolation, culture and regeneration of plantlets from tree protoplasts.

References

Ahkong, Q.F., Howell, J.I., Lucy, J.A., Safwat, F., Dawey, M.R. and Cocking, E.C. (1975) 'Fusion of hen erythrocytes with yeast protoplasts induced by polyethylene glycol', *Nature*, *255*, 66-7

Braun, A.C. (1954) 'The physiology of plant tumours', *Ann. Rev. Pt. Physiol.*, *5*, 133-52

Braun, A.C. (1974) *The Biology of Cancer*, Addison Wesley, New York

Carlson, P.S., Smith, H.H. and Dearing, R.D. (1972) 'Parasexual interspecific plant hybridisation', *Proc. Natl. Acad. Sci.* (USA), *69*, 2292-4

Carlson, P.S. (1973) 'The use of protoplasts in genetic research', *Proc. Natl. Acad. Sci.* (USA), *70*, 598-602

Cocking, E.C. (1977a) 'Protoplast fusion: progress and prospects for agriculture', *Span*, *20*, 5-8

Cocking, E.C. (1977b) 'Uptake of foreign genetic material by plant protoplasts', *Int. Rev. Cytol.*, *48*, 323-41

Chaleff, R. (1981) *Plant Somatic Genetics*, University Press, Cambridge, England

Constabel, F. (1978) 'Development of protoplast fusion products, heterokaryons, and hybrid cells' in T.A. Thorpe (ed.), *Frontiers of Plant Tissue Culture, 1978*, Calgary, pp. 141-9

Constabel, F. and Kao, K.N. (1974) 'Agglutination and fusion of plant protoplasts by polyethylene glycol', *Can. J. Bot.*, *52*, 1603-6

Davey, M.R. and Cocking, E.C. (1972) 'Uptake of bacteria by isolated higher plant protoplasts', *Nature*, *239*, 455-6

David, A. and David, H. (1979) 'Isolation and callus formation from cotyledon protoplasts of pine (*Pinus pinaster*)', *Z. Pflanzenphysiol.*, *94*, 173-7

David, H., David, A. and Mateille, T. (1982) 'Evaluation of parameters affecting the yield, viability and cell division of *Pinus pinaster* protoplasts', *Physiol. Plant*, *56*, 108-13

Gleba, Y.Y. (1978) 'Microdroplet culture: tobacco plants from single mesophyll protoplasts', *Naturwissenschaften*, *65*, 158-9

Kao, K.N. and Michayluk, M.R. (1975) 'Nutrient requirements for growth of *Vicia hajastana* cells and protoplasts at a very low population density in liquid media', *Planta*, *126*, 105-10

Kameya, T. and Uchimiya, H. (1972) 'Embryoids derived from isolated protoplasts of carrot', *Planta*, *103*, 356-60

Kirby, E.G. and Cheng, T.Y. (1979) 'Colony formation from protoplasts derived from Douglas fir cotyledons', *Plan. Sci. Lett.*, *14*, 145-54

Lörz, H., Potrykus, I. and Thomas, E. (1977) 'Somatic embryogenesis from tobacco protoplasts', *Naturwissenschaften*, *64*, 439-40

Lörz, H. and Potrykus, I. (1979) 'Regeneration of plants from mesophyll protoplasts of *Atropa belladonna*', *Experimentia*, *35*, 313-14

Mastrangelo, I.A. (1979) 'Protoplast fusion and organelle transfer' in *Nicotiana Procedures for experimental use*, United States Department of Agriculture, 65-73

Melchers, G. and Labib, G. (1974) 'Somatic hybridisation of plants fusion of protoplasts I. Selection of light resistant hybrids of 'haploid' light sensitive varieties of tobacco', *Mol. Gen. Genet.*, *135*, 277-94

Melchers, G. (1982) 'Protoplasts fusion: A review', *Proceedings 5th International Plant Tissue Culture Congress*, Japan 1982

Nagata, T. and Takebe, I. (1971) 'Plating of isolated tobacco mesophyll protoplasts on agar medium', *Planta*, *99*, 12-20

Patnaik, G., Cocking, E.C., Hamill, J. and Pental, D. (1982) 'A simple procedure for the manual isolation and identification of plant heterokaryons', *Plant Sci. Lett.*, *24*, 105-10

Potrykus, I., Harms, C.T. and Lörz, H. (1979) 'Multiple drop array (MDA) technique for the large scale testing of culture media variations in hanging microdrop cultures of single cell systems', *Plant Sci. Lett.*, *14*, 231-5

Power, J.B. and Cocking, E.C. (1971) 'Fusion of plant protoplasts', *Sci. Prog.*, *59*, 181-98

Power, J.B., Cummins, S.E. and Cocking, E.C. (1970) 'Fusion of isolated plant protoplasts', *Nature*, *225*, 1016-18

Ream, L.W. and Gordon, M.P. (1982) 'Crown gall disease and prospects for genetic manipulation of plants', *Science*, *218*, 854-9

Redenbaugh, K., Ruzin, S., Bartholomew, J. and Bassham, J.A. (1982) 'Characterization and separation of plant protoplasts via flow cytometry and cell sorting', *Z. Pflanzenphysiol.*, *107*, 65-80

Smith, H.H., Kao, K.N. and Combatti, N.C. (1976) 'Interspecific hybridisation by protoplast fusion in *Nicotiana*: confirmation and extension', *J. Hered.*, *67*, 123-8

Vasil, V. and Vasil, I.K. (1973) 'Growth and cell division in isolated plant protoplasts in microchambers' in ***Protoplasts et fusioncellules somatiques végétales***, pp. 139-49, *Coll. Int. Centre Nat. Res.*

Willison, J.H.M. and Cocking, E.C. (1972) 'The production of microfibrils at the surface of isolated tomato fruit protoplasts', ***Protoplasma***, *75*, 397-403

Winston, L.L., Parham, R.A. and Kawstinen, H.M. (1975) 'Isolation of conifer protoplasts', Rept. No. 20 ***Institute of Paper Chemistry***, Appleton, Wisconsin

10 TISSUE CULTURE CONSERVATION OF WOODY SPECIES

Christopher P. Wilkins and John H. Dodds

Over the last two decades, there has developed an increasing awareness of the necessity for the conservation of plant genetic resources. In the early 1960s, it became apparent that much of the priceless genetic diversity present both in primitive crop varieties and in related wild populations, was fast disappearing, due to the introduction of new, high-yielding cultivars. It was quickly realised that the spread of these highly-selected, uniform cultivars, together with new developments in agriculture, was threatening the existence of complex populations of primitive crop varieties. Such populations had evolved through cultivation over long periods of time and long exposure to environmental stresses, and by competing and introgressing with associated wild and weed species. Such processes of genetic erosion were also occurring in developing countries as a result of land clearance schemes for subsequent crop cultivation or urban development, and through the denudation of forests to provide timber for export or home use.

In 1964, the International Biological Programme (IBP) of the International Council of Scientific Unions (ICSU) was initiated, and a subcommittee was set up to study ways and means of collecting and conserving plant genetic resources which were threatened by agricultural developments in many of the 'centres of plant genetic diversity', first described by N.I. Vavilov nearly 40 years previously. The following year, there occurred an integration of effort between the IBP and the Food and Agriculture Organisation of the United Nations (FAO) into the problems of conservation of plant genetic resources. This collaboration resulted in two joint FAO/IBP Technical Conferences being held in 1967 and 1973. The proceedings of these conferences have been published as part of a series of IBP handbooks (1970, 1975).

Many aspects of plant genetic conservation were discussed at these conferences, including exploration, evaluation, conservation and storage, utilisation, documentation/information management and technical aspects of genetic resources centres. In addition, the 1975 conference paid particular attention to methods of conservation and storage. A comprehensive review by Roberts (1975) of the principles and methods of seed and pollen storage demonstrated that long-term seed storage was a relatively simple and inexpensive operation in terms of technology, facilities, staff and operating expenses. It was evident that seeds of most crop species could be dried to a low moisture content (5 per cent), then stored at low temperatures (-20°C) for many years with a minimum risk of genetic damage. Such seeds were termed 'orthodox'. However, it was apparent that certain categories of crops could not be conserved by such methods, and several authors proposed the use of tissue culture techniques as a possible solution to this problem (Henshaw, 1975; Morel, 1975).

There are five distinct kinds of germplasm material which are in need of preservation: cultivars in current use; obsolete cultivars; special genetic stocks such as resistance stocks, genetic and cytogenetic material, induced mutations, etc.; primitive varieties or old land races; and wild and weed species related to cultivated species. Within these five groups, three categories of crop type may be distinguished:

(1) Annual, mostly inbred, seed propagated crops with 'orthodox' seed which can be dried and stored at low temperature.
(2) Woody, perennial outbreeders that are not, or cannot yet be cloned, with characteristically short-lived seed (termed 'recalcitrant') which cannot yet be stored.
(3) Clonally propagated outbreeders, herbaceous or woody, which are highly heterozygous, and for which seed propagation is either undesirable or impossible.

It will be appreciated that those crop plants not amenable to conservation via seed storage fall into categories (2) and (3) and, as such include practically all economically important woody species. Although some progress has recently been made concerning the problems of recalcitrant seed storage, it is evident that all clonally propagated woody species must be stored in a vegetative form. Hawkes (1982) has discussed the use of conventional methods for the conservation of such species. These include the establishment of *in situ* biosphere reserves, or nature reserves, 'natural mass reservoirs', gene parks, etc., which would preserve areas of natural vegetation in their original habitat together with the species which they contain. However, such biosphere reserves have limited application since they could not hope to conserve the amount of genetic diversity which it would be desirable to preserve. The genetic diversity contained within a wild species can be partitioned into three main categories (Hawkes, 1982): (a) geographical or climal diversity, related to broad climatic and altitudinal differences; (b) interpopulation diversity in each particular sampling area; and (c) intrapopulation diversity, or the diversity to be found in each local gene pool.

Since such *in situ* biosphere reserves must of necessity be small in comparison to the total distribution of a species, such reserves would only be of use in conserving intra- and interpopulation diversity. For a widespread species, as most woody species are, they could not possibly be numerous enough to effectively conserve geographical variation. Further, even if *in situ* reserves could be established, they would be constantly under threat from changing government policies, timber needs and the rapid spread of urban developments.

An alternative, and perhaps more realistic, solution to the preservation of long-lived woody perennial species, is the establishment of plantations and fruit-tree orchards. These could be of use in the tropics, where land space is readily available and labour costs are low. However, trained personnel would still be required, and such plantations would be constantly at risk from natural disasters (earthquakes, floods, volcanos, etc.) and war. Also, there would be constant exposure to pests and pathogens, and threats from changing government policies and urban developments. Perhaps the major difficulty with this type of conservation, is the number of individuals required to provide an adequate sample of genetic diversity. Burley and Namkoong (1980) have estimated that the number of individuals needed to provide an adequate sample of the genetic diversity of a tree species would vary from 20-30 for a small single population, up to several hundred for gene pool conservation and upwards of 5000 for maintenance of heterozygosity. The preservation of trees on such a scale would involve vast areas of land. There is therefore an obvious need to develop alternative methods whereby long-lived woody perennial species may be conserved.

Since the joint FAO/IBP Technical Conference of 1974, there have been rapid advances in the application of tissue culture techniques to the problems of plant genetic conservation. For some species, e.g. potatoes (*Solanum* spp.) and cassava (*Manihot esculentum*), such techniques are already in use at International Genetics Resources Centres such as Centre Internacional de la Papa (CIP) in Peru, and at the International Centre for Tropical Agriculture (CIAT) in Colombia (Wilkins and Dodds, 1982; Withers, 1982). However, for most woody species the application of such techniques is still mostly speculative, although a recent Technical Report produced by the Secretariat of the International Board for Plant Genetic Resources (IBPGR) listed many woody species which were presently being investigated with a view to conservation via *in vitro* techniques (Withers, 1982).

Possible Applications of Tissue-culture Methods of Conservation

There are a great variety of economically important woody species which are presently undergoing genetic erosion. For some of these tree crops, notably certain members of the temperate zone tree-fruits, considerable progress has already been made regarding species conservation. However, other species, especially some tropical fruits and timbers, are already threatened with extinction.

Considering first the temperate zone tree-fruits, the main centres of genetic diversity of these species were first described by early workers such as Vavilov (1926, 1930) and Zielinski (1955). More recently, Zagaja (1970) discussed the vast range of temperate zone tree-fruits to be found in just one area of genetic diversity, namely Turkey. Surveys had revealed a vast range of genetic variation, including adaptations to specific climatic conditions such as extreme frost-resistance in certain apple species, pathogen-free populations of sweet cherries, variation in flowering period among almonds, dwarf varieties of *Prunus* species and adaptations to highly calcareous soils. However, the author emphasised the point that all such tree fruits in Turkey were threatened with extinction, due mainly to modernisation of fruit production within the country. For example, apricots which were previously propagated by seed were being exclusively propagated by grafting. Introduced peach cultivars had become so successful as to have almost entirely eliminated local varieties, and recent introductions of standard apple, pear and cherry cultivars were having a similar effect.

As well as conserving ancient genotypes, locally selected cultivars and wild species of such trees, there is also a need to conserve current and recently obsolete cultivars of fruit-tree scion varieties, and highly selected rootstock clones. Since some of these rootstock clones are sterile, i.e. the semi-dwarfing hybrid cherry rootstock Colt (*Prunus avium* x *pseudocerasus*), storage in a vegetative form is an absolute necessity.

As previously mentioned, there are numerous problems associated with the conservation of such fruit trees in orchards. These include space considerations, and the need to effectively sample and conserve as much as possible of the genetic variation contained within a population. For example, Sykes (1975) has pointed out that even sampling one tree in every 1000 of the three million almond trees in Turkey would provide 3000 single tree accessions, and would require an area of some 15 hectares. Similarly, the total area required for a germplasm collection of wild species, ecotypes and cultivars of pistachio has been estimated at 21 hectares (Maggs, 1973). It has been suggested that storage area could possibly be reduced by the use of dwarfing rootstocks or multiple grafting techniques. A recent report by Populer (1975), concerning the preservation of old apple and pear cultivars in Belgium, has indicated that keeping trees as upright cordons at 0.5 x 2.0 metre spacing would allow the maintenance of 1000 cultivars, with two trees per cultivar on 0.2 hectares. However, the same report stresses the susceptibility of such collections to pathogen attack and genetic erosion, and emphasises the need to develop suitable tissue-culture storage systems as soon as possible.

In spite of the high cost of maintaining such collections, many National and International fruit-tree germplasm repositories have been established (Brooks and Barton, 1977) and the inherent value of such collections has been amply stressed (Thompson, 1981). Furthermore, a comprehensive inventory of available North American and European fruit and tree-nut germplasm resources has recently been compiled (Fogle and Winters, 1980).

As discussed elsewhere in this volume, much progress has been made in recent years concerning the *in vitro* propagation of temperate fruit trees, and some of these techniques have already found commercial application for the rapid clonal multiplication of desirable cultivars (Jones, 1979). In addition, as previously reported (Wilkins and Dodds, 1982), the development of reliable *in vitro* methods of storage for such species is already in progress, and some further aspects of this work will be discussed in a later section.

There are a great many other woody species whose remaining genetic resources need to be urgently collected and conserved, and which would benfit greatly from the development of reliable *in vitro* methods of both propagation and conservation. Tisserat, Ulrich and Finkle, (1981) have reported that date palm (*Phoenix dactylifera*) genomes in Algeria and Morocco are bordering on extinction due to Bayond disease (*Fusarium oxysporium* Schlect. var. *albedinis*). However, conservation by conventional means would be virtually impossible since there is no efficient method of vegetative propagation. The conventional method of reproduction of date-palm, i.e. via suckers arising from the base of the main stem, is both slow and unreliable. Fortunately, a solution to this problem now exists, since reliable methods for both *in vitro* propagation and cyopreservation of date palm have recently been developed (Ulrich, Finkle and Tisserat, 1982). Some additional aspects of this work will be discussed in a later section. Similar systems of *in vitro* propagation and conservation are also being developed for other palm species such as oil palm (*Elaeis guineensis*). Corley, Barrett and Jones (1976) have reported the introduction of tissue cultured oil palms into free-living plantation conditions, whilst Withers (1982) has reported the cyopreservation of zygotic embryos of oil palm with high levels (80 per cent) of success. However, in the case of the coconut palm (*Cocos nucifera*) little progress has so far been made, and the successful regeneration of coconut plantlets from an *in vitro* system has not yet been reported. In addition, there is an urgent need for programmes to be initiated for the intensive collection of coconut germplasm throughout South East Asia and the Pacific, due to the rapid loss of genetic variability and the need to incorporate a wider genetic base into breeding work (Williams, 1982). However, little attention has so far been given to this urgent task, due to the cost and practical problems involved in sampling and transporting large quantities of nuts.

Such problems have also prevented the collection of dwindling germplasm resources of other crops such as cocoa (*Theobroma cocao*). Cocoa scientists have recently (Williams, 1982) agreed on an urgent need to collect throughout Central America for *criollo* varieties, and in the whole of the Amazon basin for *amazonia* types. In particular, they noted the extinction of variability which was known to exist in 1973.

In spite of these discouraging reports, much progress has recently been made regarding the establishment of germplasm repositories for certain tropical fruits. Coronel (1982) has recently discussed the establishment of an extensive collection of tropical fruit species in the Philippines. However, even if collections such as these could be established for the crops previously discussed, the safety of such collections cannot be guaranteed. There is therefore an urgent need to develop reliable methods of *in vitro* propagation and conservation for these species.

Perhaps the group of species which are most threatened at the present time are the luxury timber crops. A recent publication by the American National Academy of Sciences (1979) has outlined some of the urgent research needs of these species. One such threatened species is the tropical hardwood known as afrormosia (*Pericopsis elata*) or African teak. This species, native to various West African countries, provides one of the most valuable woods on world timber markets. As a consequence, most natural stands of this tree have been heavily cut. Since the natural regeneration of afrormosia is negligible and trees are not being planted on a large scale, the species is facing economic and biological extinction. Immediate action must therefore be taken in order to preserve afrormosia germplasm, and it has been recommended that Ghana, Cameroon and Zaire each set aside 2-3 km^2 of afrormosia forest as a conservation reserve, protected from exploitation, but available to foresters for germplasm collection. However, this method of conservation has obvious drawbacks, and in addition, afrormosia seeds have a maximum storage life of only a few months. The development of reliable systems of *in vitro* propagation and storage would facilitate the preservation of what remains of afrormosia germplasm, and would also aid re-forestation programmes via rapid clonal multiplication of elite trees. There is also a dire need for research dealing with fundamental aspects of afrormosia silviculture. Apart from the fact that afrormosia seedlings require open sunlight and therefore will not grow beneath an already established canopy, foresters have virtually no knowledge of the flowering and seeding habits of the species, including: pollination methods, fruit ripening and frequency of seed years, seed germination, ecological requirements of young seedlings and the performance of trees under plantation conditions.

Similar proposals have also been made for various *Intsia* species, in an attempt to conserve remaining stocks. Members of this valuable genus have been so intensively exploited throughout much of South East Asia, that in most countries few trees are left in natural stands. Some countries (i.e. Malaysia) have already restricted the export of *Intsia* wood, but the complete decline of these species as economic plants is imminent. Again, as for afrormosia, little is known of these species regarding environmental tolerances, seeding and flowering habits and susceptibilities to pests and pathogens. Urgent research is needed in the following areas: seed germination, seedling characteristics, nodulation, nursery requirements, transplantation and basic silvicultural requirements such as shade tolerance, optimum thinning time and planting density, and the need for mixing with other species. There are therefore obvious arguments in favour of the development of reliable methods of *in vitro* propagation and storage.

Similar concern has also been expressed regarding various members of the two genera *Pterocarpus* and *Dalbergia*. Species of *Pterocarpus* include the timbers padauk, narra and muninga – some of the most valuable on International trade markets. None of these species is extensively cultivated and native stands are fast disappearing; such species are also untried in plantations. However, a recent report by Rao and Lee (1982) has indicated that some progress is now being made regarding the tissue-culture propagation of these species. Hopefully, these techniques will eventually provide the basis of an *in vitro* system of genetic conservation. A similar situation is apparent in the case of the genus *Dalbergia*, some dozen or so species of which comprise the rosewoods. All accessible stands of these species are now gone and, apart from in Java and India, there are no plantations or trial plantings. However, a recent encouraging report by Mascarenhas *et al.* (1982) has indicated that a system may soon be available for the rapid *in vitro* multiplication of these species, and studies aimed at *in vitro* storage could then commence.

As well as endangered species of tropical timber trees, there are many tropical fruits for which genetic conservation is urgently required. In the case of tamarind (*Tamarindus indica*), a tropical fruit whose possible potential is just beginning to be realised, germplasm collections need to be made in a region that stretches from Senegal across sub-Sahelian Africa to the Sudan, and also in India, Thailand and South-East Asia. Other tropical fruits which could benefit enormously from the application of *in vitro* techniques include the mangosteen (*Garcinia mangostana*). This fruit is unavailable in what could be its major markets – Central America, South America, Australia and Africa – since there is no reliable method for vegetative propagation.

The *in vitro* Culture of Woody Plants

Since this volume is a testimony to the present 'state of the art' regarding the *in vitro* culture of woody plants, any additional discussion of this subject would be inappropriate. This section will therefore briefly reiterate the main benefits of tissue culture systems and then discuss the intrinsic limitations of such systems when applied to the problems of plant genetic conservation.

The main advantages of tissue culture systems can be listed as follows:

(1) Most *in vitro* systems possess the potential for very high multiplication rates: for example, Morel (1975) has reported that with grapes it is possible to obtain several million plants in the space of one year from a single meristem explant. Similarly, by means of shoot-tip culture of apple, it is possible to obtain several tens of thousands of clonal plants per year (Jones, Pontikis and Hopgood, 1979).
(2) Tissue culture systems are aseptic, and can be easily kept free from

fungi, bacteria, viruses and insect pests. One important use of *in vitro* techniques is the production of pathogen-free stocks for use by plant breeders.

(3) Space considerations: Galzy (1969) has demonstrated that by storing plantlets of grape (*Vitis rupestris*) at a low temperature (9°C) for 1 year, a total of 800 cultivars with 6 replicates per cultivar can be stored in an area of 2 m^2. The comparable *in vivo* situation would have required 1 hectare of field space.

(4) In an ideal tissue-culture storage system, genetic erosion is reduced to zero.

(5) By means of specialised *in vitro* techniques, i.e. pollen and anther culture, haploid plants may be produced which are of use in breeding programmes.

(6) Tissue culture conservation techniques are also useful in plant breeding programmes as a means of rescuing and subsequently culturing zygotic embryos from incompatible crosses which normally result in embryo abscission. Such a technique has been successfully applied to avocado (Sedgley and Alexander, 1980).

(7) The expense, both in labour and financial terms of maintaining large collections of plants in the field can be greatly reduced by the use of efficient tissue-culture storage techniques.

A basic pre-requisite of any tissue culture system which is to be employed for the purpose of genetic conservation, is the ability to regenerate genetically stable independently growing plants from such a system with a high level of success. Such a criterion normally dictates the use of organised cultures such as those derived from shoot apical meristems or shoot-tips, and in which further proliferation of shoots is achieved by inducing the outgrowth of pre-existing meristems. Culture systems dependent on adventitious shoot development or somatic embryogenesis from either callus cultures or cell suspensions are usually considered to be inherently genetically unstable.

Such genetic instability may derive from instability within the initial explant source, or arise as a result of subsequent culture conditions. It has been suggested that more effort be directed towards establishing causal relationships between storage/culture conditions and the generation of instability in *in vitro* cultures (Henshaw, 1982). In spite of the inherent problems involved in the production of plants via *in vitro* somatic embryogenesis from callus cultures, for species in which the branching habit is infrequent or non-existent, such as certain palms, there is no alternative, since propagation through shoot-tip cultures is not possible. Many authors have discussed and reviewed the problems of genetic instability in plant cell, tissue and organ cultures (D'Amato, 1975, 1978; D'Amato *et al.*, 1980).

There are many woody species for which systems of *in vitro* propagation have not yet been elucidated. For various species, problems have been encountered at all stages of culture, including: culture initiation, culture establishment, achieving satisfactory rates of shoot proliferation or embryoid formation, and especially the *in vitro* induction of rooting and the subsequent establishment of rooted plantlets in free-living conditions. The phenomenon of explant blackening also appears to be especially prevalent during the *in vitro* culture of woody species. A recent review by Withers (1982) also listed many problems experienced during *in vitro* conservation work, these include: lack of equipment, shortage of experienced personnel, shortage of fresh plant material or appropriate cultivars, contamination problems and lack of funds. In addition to these problems, there also appears to be some reticence towards the actual application of *in vitro* techniques for germplasm storage. Williams (1982) has recently cited the recommendations of grape breeders that germplasm collections should be made in the Near and Middle East, North Africa, parts of the Himalayas and North East China, but that no progress has so far been made due to the problems of planting and maintaining huge collections. However, *in vitro* storage of grape plantlets has been feasible for many years (Galzy, 1969; Morel, 1975).

Methods of in vitro *Conservation*

There are three possible approaches to the storage of woody plants via *in vitro* techniques: storage by continued growth of cultures at 'normal' rates of growth involving a continuous cycle of transferring cultures to fresh culture media at regular intervals; storage of cultures under conditions of minimal growth; total suspension of all growth and metabolic processes by ultra-low temperature storage (cryopreservation). Some relevant aspects of each of these methods will be discussed in turn.

Normal Growth Storage. It will be evident from discussions in previous sections that an *in vitro* system with a high multiplication rate, although ideal for purposes of clonal propagation, is unsuitable as a means of germplasm conservation. Such systems require frequent attention and maintenance, and the mutation rate is likely to correspond directly to the rate of cell division. However, such a method of storage may be useful for maintaining shoot cultures of species for which regeneration procedures (i.e. root induction and subsequent plantlet establishment) have not yet been fully elucidated. Also, maintaining cultures in this fashion may prove useful during investigations designed to evaluate the most efficient method of growth suppression, or as part of a cryopreservation programme. This system has been used at Birmingham to maintain germplasm stocks of various *Malus* species which are not usually cultivated on a commercial basis, such as *M. prunifolia* and *M. baccata* (Siberian crab apple) (Figure 10.1a, b).

Minimal Growth Storage. Techniques of germplasm conservation based on the storage of shoot-tip cultures, or meristem-derived plantlets under conditions which permit only minimal rates of growth are likely to have widespread application in the near future. Such methods of storage possess the advantages that the stored material is readily available for use, can be easily seen to be alive, and cultures may be easily replenished when necessary. There are several possible approaches when attempting to store tissue-cultures via growth suppression, these can usually be divided into three categories:

(1) Altering the physical conditions of culture. This may include temperature or the gaseous conditions (the partial pressure of oxygen) within the culture vessel.

(2) Alterations to the basic culture medium, i.e. a sub- or super-normal nutrient availability, or the omission/reduction of some factor which is usually essential for normal growth.

(3) Addition of growth retardants such as abscisic acid (ABA), chlorophonum chloride (Phosphon D), maleic hydrazide, diaminiazide (B995), cycocel, (2-chloroethyl)-trimethyl ammonium chloride (CCC), N-dimethyl-succinamic acid and ancymidol, or compounds with osmotic effect such as the sugar-alcohols mannitol and sorbitol.

Temperature reduction as a means of storage of *in vitro* cultures has been used to store meristem-derived plantlets of grape (*Vitis rupestris*) and shoots of the apple scion cultivar 'Golden Delicious' (*Malus domestica*). Such a technique also allows efficient international exchange of pathogen-free germplasm stocks *in vitro*, without risk of spreading diseases, and also avoiding quarantine regulations.

Work at Birmingham has shown that cultures of temperate fruit-trees may be stored for many months by the use of several different techniques. The survival of cherry (*Prunus avium* x *pseudocerasus*) shoot cultures at normal culture temperatures for several months, when grown on liquid medium, has already been reported (Wilkins and Dodds, 1982). Similar results have now been obtained for various apple, plum

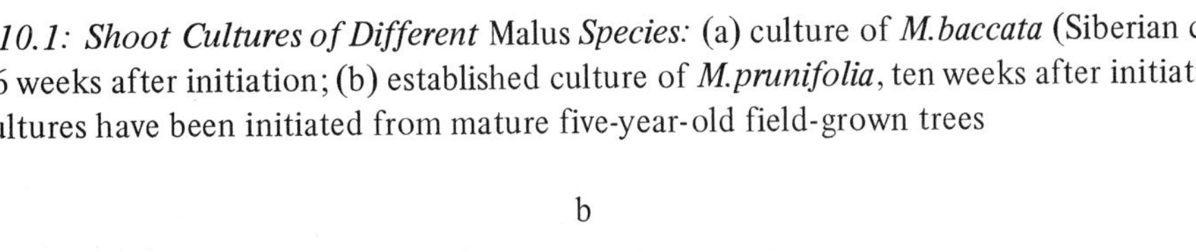

Figure 10.1: Shoot Cultures of Different Malus *Species:* (a) culture of *M.baccata* (Siberian crab apple) 6 weeks after initiation; (b) established culture of *M.prunifolia*, ten weeks after initiation. Both cultures have been initiated from mature five-year-old field-grown trees

a

b

Figure 10.2: Maintenance of Fruit-tree Cultures at Normal Culture Temperatures: (a) proliferating shoot culture of the plum rootstock Pixy (*Prunus insititia*), after twelve months growth on liquid medium at 23°C; (b) jar containing single shoots of the apple rootstock M7 (*Malus domestica*) after nine months incubation at 25°C. Shoots are growing on 50 ml of agar-solidified medium containing a greatly reduced concentration of growth regulator

a

b

and pear cultivars. By means of this system, shoot cultures of the plum rootstock Pixy (*Prunus insititia*) were kept for 12 months at 25°C (Figure 10.2a). After this storage period, from 3-4 shoots per jar could be transferred onto fresh proliferation medium and healthy proliferating cultures were rapidly established. One of the major disadvantages of growing fruit-tree cultures on agar medium is that browning of the medium occurs if sub-culturing is delayed. This is usually followed by rapid senescence of the culture; however this phenomenon does not occur when cultures are grown on liquid medium. Preliminary results indicate that single shoot 'microcuttings' of apple cultivars may be stored for many months on agar medium with a greatly reduced concentration of growth regulators (Figure 10.2b). In addition, this technique gives rise to extended shoots with several nodes, each of which may be used to initiate fresh cultures. The use of jars containing fairly large volumes of culture medium (50-100 ml) avoids problems of nutrient depletion when shoots are stored at normal temperatures, although single tubes are preferable in terms of space requirements and avoiding contamination problems.

Low temperature (4°C) storage has been successfully employed to store shoots of various apple rootstock (Figure 10.3) and scion cultivars for many months, with high levels of success. The criterion for survival of cultures is taken as the presence of shoots of normal appearance, which continue growth and form proliferating cultures when transferred to fresh medium at normal culture temperatures. Other systems of *in vitro* storage of temperate fruit trees which are being investigated include: (a) storage of single shoots on solid medium (Figure 10.4a) with greatly reduced concentrations of growth regulators, but at normal temperatures; (b) storage of single shoots at reduced temperature (4°C) on solid medium with normal concentrations of growth regulators required for proliferation (this system of storage may be desirable for purposes of germplasm exchange); or (c) storage of rooted plantlets (Figure 10.4b) on liquid medium with a filter paper strip as support. System (a) would require the transfer of the stored shoot to either rooting medium (to obtain plantlets), or proliferation medium (to obtain further shoots) after the appropriate period of storage. System (b) allows the production of further shoots when desired by simply returning the stored shoot to normal culture conditions. System (c) provides a supply of readily available rooted plantlets which may be simply removed from storage and established in the glasshouse. Investigations into the use of various growth retardants as a means of minimal growth storage are also in progress. There have been few other reports in the literature regarding the use of minimal growth storage for conservation of woody species. Kartha *et al.* (1981) have successfully stored shoots of coffee by minimal growth techniques, and such methods of storage are also in prospect for cultures of forest tree species such as *Pseudotsuga menziesii* and *Sequoia sempervirons* (Withers, 1982).

Figure 10.3: Single Shoots of Various Apple Rootstocks (Malus domestica) *Stored for Twelve Months on Liquid Medium at 4°C:* (a) M25, (b) M26 and (c) M27

a b c

Freeze-preservation. It will be evident that an ideal system for germplasm storage of woody species would be to store material in such a manner as to achieve a complete cessation of cell division. This may be achieved by storage at the temperature of liquid nitrogen (-196°C). A comprehensive review by Withers (1980a) and references cited therein, has discussed fully all aspects of the cryopreservation of plant cell, tissue and organ cultures. Only the basic principles and methodology of freeze-preservation will be discussed in the following section.

Figure 10.4: Examples of the Types of Culture System Used for Minimal Growth Storage of Temperate Fruit Trees: (a) single shoot of the apple rootstock M27 (*Malus domestica*), stored on agar solidified medium under conditions of minimal growth; (b) rooted plantlet of the cherry rootstock Colt (*Prunus avium* x *pseudocerasus*) stored on liquid culture medium with a filter paper strip as support. (Reproduced with permission from *Sci. Prog.* (Oxf.), *68*, 281-307, 1982)

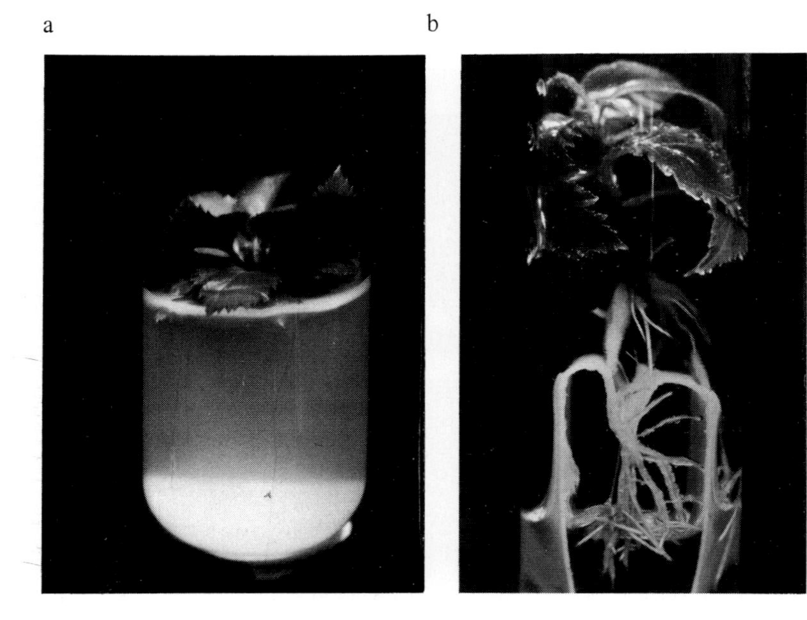

Principles of Cryopreservation

The successful freeze-preservation of a particular plant tissue requires that damage caused by ice-crystal formation within the individual cells of the tissue is either prevented or minimised. This has resulted in the adoption of two basic approaches to cryopreservation; these are ultra-rapid freezing and slow/stepwise freezing. The process of rapid freezing results in the formation of ice crystals within the cells which are of microscopic size, and which do not disrupt the internal organelles and membranes of the cells. However, thawing must be carried out rapidly enough to prevent recrystallisation. Slow or stepwise freezing has been applied to most tissue-culture systems and depends upon extracellular freezing for protection. As cells are cooled, the surrounding liquid (either intercellular fluid or extracellular culture medium) will eventually freeze due to ice nucleation. However, the intracellular fluid will not yet be frozen, and so a process of cellular dehydration will be initiated due to the water vapour pressure deficit established by extracellular ice formation. The concentration of intracellular solutes resulting from dehydration then further depresses the freezing point of the cell contents. This process of gradual removal of intracellular water is termed 'protective dehydration', and effectively prevents ice formation in the cytoplasm or vacuole. However, overdehydration can result in damage to the cell due to 'solution effects' (Merryman, Williams and Douglas, 1977). In addition, the use of a chemical cryoprotectant is normal in either of the above two freezing protocols.

Methodology of freeze-preservation

The following section describes briefly the major procedural steps which comprise the process of freeze-storage, and considers some of the materials and equipment necessary for programmes involving the freeze-preservation of plant germplasm.

Cryoprotection. Many compounds, or combinations of such compounds, have been employed as cryoprotectants. These include dimethylsulphoxide (DMSO), glycerol, proline, hydroxyethyl starch polyvinylpyrrolidone, polyethylene-glycol and various sugars. Many mechanisms have been proposed by which cryoprotectants are thought to reduce cryodamage, which have been summarised by Withers (1980a) and include an ability to reduce the size and growth rate of ice crystals, a lowering of the freezing point of the intracellular contents so facilitating protective dehydration in the early stages of freezing, stabilisation of macromolecules and membranes within the cell, a colligative action in which the effective concentration of solutes in equilibrium with ice inside and outside the cell is lowered, and a contribution to osmotic dehydration of the cell before freezing. Concentration and modes of application of cryoprotectants have been discussed by Withers (1980a) and references cited therein.

Freezing. The freezing process is usually carried out with the aid of some form of apparatus which provides a controlled application of liquid nitrogen or liquid nitrogen vapour to the specimen being frozen. Purpose-built programmable biological freezers are available from various sources, and there are many references to these devices in the

Figure 10.5: Diagrammatic Representation of a Programmable Freezing Apparatus. (Reproduced with permission from *Sci. Prog.* (Oxf.), *68*, 281-307, 1982)

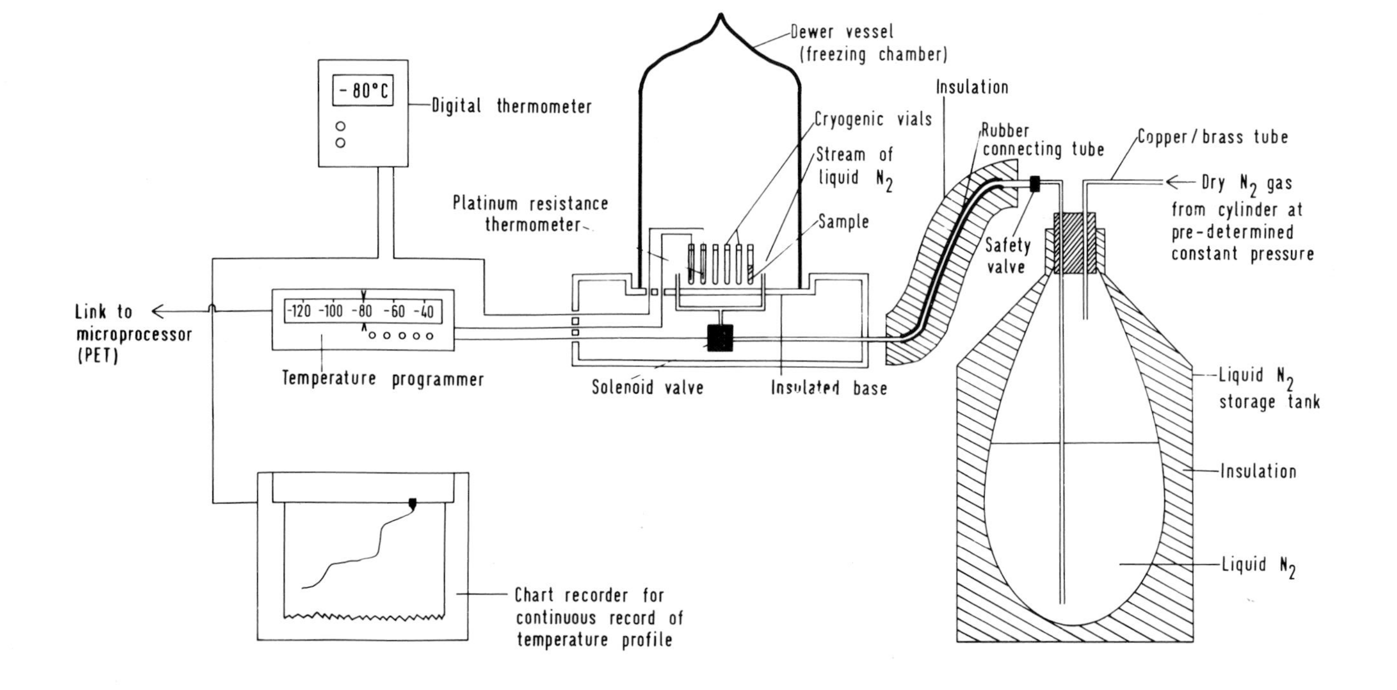

literature. However, in view of the high cost of these machines, several authors have prepared detailed instructions for the building of cheaper, easily constructed freezing devices that are capable of giving precisely controlled rates of freezing and thawing. Figure 10.5 gives details of a programmable freezing apparatus built at Birmingham, which is easily and cheaply constructed from readily-available components. Such a device will suffice for all types of freeze-thawing protocols, including fast, slow and/or stepwise freezing.

Very rapid rates of freezing may be obtained by suspending the specimen on the end of a hypodermic needle, then plunging the needle into liquid nitrogen, to a pre-determined depth.

Storage. Once frozen, specimens are usually stored in 2 ml polypropylene ampoules in a purpose-built liquid nitrogen fridge (Figure 10.6). Such fridges consist simply of insulated containers holding a large volume of liquid nitrogen, and are cheap and simple to maintain. Ampoules are normally placed into stainless-steel holders for ease of handling, then stored in open metal cans within the fridge. Some 4000 x 2 ml ampoules can be stored in a standard liquid nitrogen fridge. Figure 10.6 illustrates an optimal feature of some liquid nitrogen fridges: the ampoule holder (bottom right) is located within the top portion (bottom left) of the liquid nitrogen fridge and is therefore within the liquid nitrogen vapour phase at a temperature of around –130°C to –150°C. By adjusting the position of the plastic 'O' ring on the shaft of the ampoule holder, variable slow rates of freezing can be achieved.

Thawing. Thawing of most frozen plant tissues must usually be carried out rapidly, by exposing the specimen ampoule to warm water at 37-40°C. The reason for this is that some freezable water may remain within the cells, and if thawing is not sufficiently fast, recrystallisation may occur. The resultant ice crystals are liable to disrupt intracelluar organelles and membranes and result in the death of the cell.

Assessments of Viability and Cryoinjury. An obvious pre-requisite of any tissue-culture system of germplasm conservation is the need for a definitive criterion of survival. Although a resumption of growth of a culture is the most conclusive test of survival, such an event does not quantitatively reflect the extent of cellular survival and damage. Several viability tests have therefore been evolved to ascertain the condition of cells immediately after thawing. Two such tests are in common use; these are fluorescein diacetate (FDA) staining and triphenyl tetrazolium

Figure 10.6: Apparatus Used for Long-term Cryopreservation of Plant Tissue Cultures: Top left – liquid nitrogen fridge; top right – ampoule rack; below – apparatus used for slow freezing of specimens in vapour phase of liquid nitrogen. Adjustment of the position of the plastic 'O' ring on the shaft of the holder (lower right) provides variable rates of freezing. (Reproduced with permission from *Sci. Prog.* (Oxf.), *68*, 281-307, 1982)

chloride (TTC) reduction. Details of the experimental procedure of these tests have been given by Withers (1980b). The use of FDA staining for assessing the extent of freezing damage in a cell suspension of sycamore (*Acer pseudoplatanus*) is illustrated in Figure 10.7a. Both of these methods should be used with caution, since they may grossly overestimate 'recovery potential', and the two tests do not necessarily provide correlating data. A useful alternative method is that of Evans' blue staining. This stain is excluded by viable cells, and its use for determining the survival of thawed cells of sycamore is illustrated in Figure 10.7b.

Figure 10.7: Specific Staining for Cell Viability. A mixture of live and dead cells was stained with fluorescein diacetate and Evans' blue. Observation (a) under ultraviolet light reveals the presence of fluorescing viable cells, whilst (b) observation under bright field tungsten illumination reveals the presence of dead cells stained by Evans' blue. (Reproduced with permission of Dr L.A. Withers, from *Cryobiology*, *15*, 87, 1978)

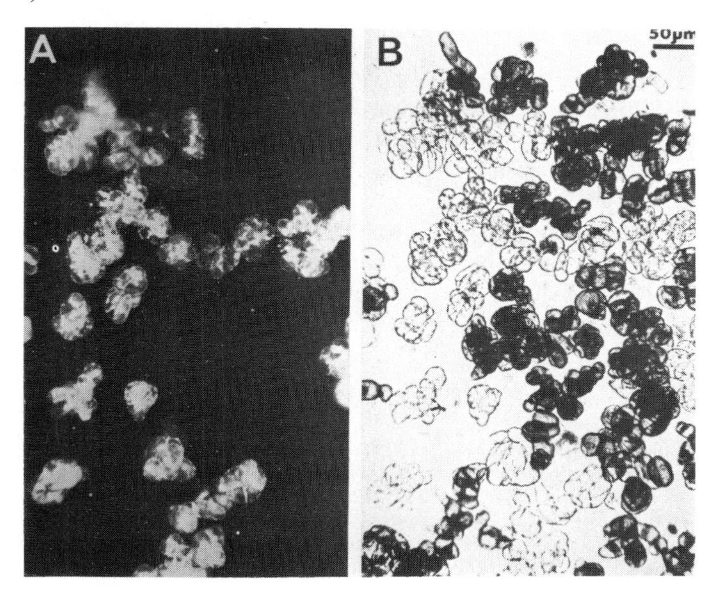

Present Applications of Cryopreservation and Future Prospects

There have been very few reports concerning the cryopreservation of *in vitro* cultures of woody plants. Sakai and Sugawara (1973) successfully cryopreserved callus of poplar (*Populus euramericana*). Withers (1978) has reported the freeze-preservation of a cell suspension of sycamore (*Acer pseudoplatanus*), whilst Simmovitch (1979) reported the cryopreservation of protoplasts isolated from the bark of black locust trees. However, none of these tissue-culture systems was morphogenic.

As yet, there have been no reports concerning the successful cryopreservation of meristems of tree species and the subsequent regeneration of free-living plants. However, a series of recent reports by Tisserat (1979), Tisserat *et al.* (1981) and Ulrich *et al.* (1982) has provided what is perhaps a perfect example of the successful application of tissue-culture technology to the problems of germplasm conservation of woody species. A method was developed for the efficient *in vitro* propagation of date palm, a species which has no reliable method of vegetative propagation, and for which germplasm resources in several parts of the world are threatened with extinction. Aspects of cryopreservation were then investigated, and a method was developed for the freeze-preservation of embryogenic date-palm callus. Studies of enzyme polymorphism in five gene-associated enzyme systems were then used as a means of detecting possible genetic alterations caused by the cryogenic treatments.

It will be evident that there is an urgent need to develop reliable methods of *in vitro* propagation and conservation for many woody species. However, in the case of the majority of the tropical tree crops, it should be stressed that perhaps a substantial part of this work should be carried out in the third world countries, from where the majority of such species originate. The work could therefore be carried out where the appropriate plant materials are most readily available, and where the local needs and practical requirements are most evident.

References

Brooks, H.J. and Barton, D.W. (1977) 'A plan for national fruit and nut germplasm repositories', *Hort. Sci.*, *12*, 298-300

Burley, J. and Namkoong, G. (1980) 'Conservation of Forest Genetic Resources', *XI Commonwealth Forestry Conference*, Trinidad

Corley, R.H.V., Barrett, J.H and Jones, L.H. (1976) 'Vegetative propagation of oil palm via tissue culture', *Malaysian International Agriculture Oil Palm Conference* 1976: 1-7

Coronel, R.E. (1982) 'Fruit Collections in the Philippines', *IBPGR Newsletter*, Regional Committee for S.E. Asia, *6*(2), 6-10

D'Amato, F. (1975) 'The problems of genetic stability in plant tissue and cell culture' in O.H. Frankel and J.G. Hawkes (eds.), *Crop Genetic Resources for Today and Tomorrow*, University Press, Cambridge, pp. 333-48

D'Amato, F. (1978) 'Chromosome number variation in cultured cells and regenerated plants' in T.A. Thorpe (ed.), *Frontiers of Plant Tissue Culture, 1978*, International Association for Plant Tissue Culture, pp. 287-96

D'Amato, F., Bennici, A., Cionini, S., Baroncelli, P.G. and Lupi, M.C. (1980) 'Nuclear fragmentation followed by mitosis as mechanism for wide chromosome number variation in tissue cultures: Its implications for plant regeneration' in F. Sala, B. Parisi, R. Cella and O. Citerri (eds.), *Plant Cell Cultures: Results and Perspectives*, Elsevier/North-Holland Biomedical Press, Amsterdam

Fogle, H.W. and Winters, H.F. (1980) *North American and European fruit and tree nut germplasm resources inventory*, U.S. Department of Agriculture, Miscellaneous Publication No. 1406, 732 pp.

Galzy, R. (1969) 'Recherches sur la croissance de *Vitis rupestris* Scheele sain et court noice cultivé *in vitro* à differentes temperatures', *Ann. Phytopathol.*, *1*, 149-66

Hawkes, J.G. (1982) 'Genetic Conservation of 'Recalcitrant Species' – An Overview' in L.A. Withers and J.T. Williams (eds.), *Crop Genetic Resources the Conservation of Difficult Material*, 1980, I.U.B.S. Series, B42, pp. 83-92

Henshaw, G.G. (1975) 'Technical aspects of tissue culture storage for genetic conservation' in O.H. Frankel and J.G. Hawkes (eds.), *Crop Genetic Resources for Today and Tomorrow*, University Press, Cambridge, pp. 349-58

Henshaw, G.G. (1982) 'Genetic Stability and Deterioration of Stored Material' in L.A. Withers and J.T. Williams (eds.), *Crop Genetic Resources the Conservation of Difficult Material*, 1980, I.U.B.S. Series, B42, pp. 97-100

I.B.P. Handbook No. 11 (1970) *Genetic Resources in Plants – their Exploration and Conservation*, O.H. Frankel and E. Bennett (eds.), Blackwell Scientific Publications, Oxford

I.B.P. Handbook No. 2 (1975) *Crop Genetic Resources for Today and Tomorrow*, O.H. Frankel and J.G. Hawkes (eds.), University Press, Cambridge

Jones, O.P. (1979) 'Propagation in vitro of apple trees and other woody fruit plants', *Sci. Hort.*, *30*, 44-8

Jones, O.P., Pontikis, C.A. and Hopgood, M.E. (1979) 'Propagation *in vitro* of five apple scion cultivars', *J. Hort. Sci.*, *54*, 155-8

Kartha, K.K., Mroginski, L.A., Pahl, K. and Leung, N.L. (1981) 'Germplasm preservation of coffee (*Coffea arabica* L.) by in vitro culture of shoot apical meristems', *Plant Sci. Lett.*, *22*, 301-7

Maggs, D.H. (1973) 'Genetic resources in pistachie', *Plant Genetic Resources Newsletter*, *29*, 7-15

Mascarenhas, A.F., Hazara, S., Potdar, U., Kulkarni, D.K., Khuspe, S.S., Paranjpe, S.V. and Gupta, P.K. (1982) 'Rapid Clonal Multiplication of Mature Forest Trees through Tissue Culture' in *Programme and Abstracts. Fifth International Congress of Plant Tissue and Cell Culture*, Japan, p. 118 (abs. H2-1)

Merryman, H.T., Williams, R.J. and Douglas, M.St.J. (1977) 'Freezing injury from solution effects and its prevention by natural or artificial cryoprotection', *Cryobiology*, *14*, 287-302

Morel, G. (1975) 'Meristem culture techniques for the long-term storage of cultivated plants' in O.H. Frankel and J.G. Hawkes (eds.), *Crop Genetic Resources for Today and Tomorrow*, University Press, Cambridge, pp. 327-32

National Academy of Sciences (USA) publication (1979) *Tropical Legumes: Re-Sources for the Future*

Populer, C. (1975) 'Old Apple and Pear Cultivars in Belgium', *Plant Genetic Resources Newsletter*, *40*, 6-7

Rao, A.N. and Lee, S.K. (1982) 'Importance of Tissue Culture in Tree Propagation' in *Programme and Abstracts. Fifth International Congress of Plant Tissue and Cell Culture*, Japan, p. 117 (abs. H2-R)

Roberts, E.H. (1975) 'Problems of long-term storage of seed and pollen for genetic resources conservation' in O.H. Frankel and J.G. Hawkes (eds.), *Crop Genetic Resources for Today and Tomorrow*, University Press, Cambridge, pp. 269-95

Sakai, A. and Sugawara, Y. (1973) 'Survival of poplar callus at super low temperatures after cold acclimation', *Plant and Cell Physiol.*, *14*, 1201-4

Sedgley, M. and Alexander, D.M.E. (1980) 'Avocado Breeding', *Proc. Int. Plant. Prop. Soc.*, *30*, 1-36

Simmovitch, D. (1978) 'Protoplasts surviving freezing to −196°C and osmotic dehydration in 5 molar salt solutions prepared from the bark of winter black locust trees', *Plant Physiol.*, *63*, 722-5

Sykes, J.T. (1975) 'Tree crops' in O.H. Frankel and J.G. Hawkes (eds.), *Crop Genetic Resources for Today and Tomorrow*, University Press, Cambridge, pp. 123-37

Thompson, M.M. (1981) 'Utilisation of fruit and nut germplasm', *Hort. Sci.*, *16*(2), 6-9

Tisserat, B. (1979) 'Propagation of date palm (*Phoenix dactylifera* L.) *in vitro*', *J. Exp. Bot.*, *30*, 1275-83

Tisserat, B., Ulrich, J.M. and Finkle, B.J. (1981) 'Cryogenic preservation and regeneration of date palm tissue', *Hort. Sci.*, *16*, 47-8

Ulrich, J.M., Finkle, B.J. and Tisserat, B.H. (1982) 'Effects of Cryogenic Treatment on Plantlet Production from Frozen and Unfrozen Date Palm Callus', *Plant Physiol.*, *69*, 624-7

Vavilov, N.I. (1926) *Studies on the Origin of Cultivated Plants*, Leningrad

Vavilov, N.I. (1930) 'Wild progenitors of the fruit trees of Turkistan and the Caucasus and the problem of origin of fruit trees', *Proc. 9th Int. Hort. Congr.*, London, pp. 271-8

Wilkins, C.P. and Dodds, J.H. (1982) 'The application of tissue culture techniques to plant genetic conservation', *Sci. Prog.* (Oxf.), *68*, 281-307

Williams, J.T. (1982) 'Techniques for Genetic Conservation: The Way Ahead' in L.A. Withers and J.T. Williams (eds.), *Crop Genetic Resources the Conservation of Difficult Material*, I.U.B.S. Series (1980), pp. 115-18

Withers, L.A. (1978) 'The freeze-preservation of synchronously dividing cultured cells of *Acer pseudoplatanus* L.', *Cryobiology*, *15*, 87-92

Withers, L.A. (1980a) *Tissue Culture Storage for Genetic Conservation*, IBPGR Technical Report, IBPGR Secretariat, Rome

Withers, L.A. (1980b) 'Cryopreservation of plant cell and tissue cultures' in D. Ingram and J. Helgeson (eds.), *Tissue Culture for Plant Pathologists*, Blackwell Scientific Publications, Oxford

Withers, L.A. (1982) *Institutes Working on Tissue Culture for Genetic Conservation*, IBPGR Technical Report, IBPGR Secretariat, Rome

Zagaja, S.W. (1970) 'Temperate zone tree fruits' in O.H. Frankel and E. Bennett (eds.), *Genetic Resources in Plants – their Exploration and Conservation*, IBP Handbook, No. 11, pp. 327-33, Blackwell Scientific Publications, Oxford

Zielinski, Q.B. (1955) *Modern Systematic Pomology*, Iowa State University Press, Ames, Iowa

11 CONCLUSIONS

John H. Dodds

As has been outlined in the introduction and several other chapters of this book, there is now an excellent range of techniques available within the field of plant tissue culture. Many of the techniques (e.g. callus culture and micropropagation) are now very well developed and have become a routine part of horticultural and agricultural practice. Some of the techniques (e.g. somatic hybridisation and genetic engineering) are still in their infancy and a great amount of research will be required before they reach any meaningful, applicable results.

One of the key features of plant tissue, cell and organ culture is that it has tended to concentrate very heavily on a few specific members of the plant kingdom. A number of paradigms or model systems have been developed to obtain a greater academic knowledge of the science, without expanding the knowledge tc a wide range of plants. Many of the important crop plants (e.g. cereals, beans) are now being studied intensely, but still relatively few studies have been done on woody material.

It is hoped that this book may indicate some of the groundwork and potential of the study of tissue culture of trees and will stimulate further study of the *in vitro* culture of woody plants. The economic importance of trees both in terms of fruit crops and timber and paper production is well known; it is most important therefore that we should employ the development of new technologies to try and improve the quality and quantity of trees produced on a global scale. Although the prospects for significant application of these new technologies to trees is a long-term one, it cannot be doubted that the rewards will be very significant.

AUTHOR INDEX

SUBJECT INDEX